PAGES OF STONE

GEOLOGY OF THE GRAND CANYON & PLATEAU COUNTRY NATIONAL PARKS & MONUMENTS

SECOND EDITION

PAGES OF STONE

GEOLOGY OF THE GRAND CANYON & PLATEAU COUNTRY NATIONAL PARKS & MONUMENTS

SECOND EDITION

HALKA CHRONIC & LUCY CHRONIC

THE MOUNTAINEERS BOOKS

In fond remembrance of Eddie McKee, Barbara McKee, and Tad Nichols

THE MOUNTAINEERS BOOKS
is the nonprofit publishing arm of The Mountaineers Club, an organization founded in 1906 and dedicated to the exploration, preservation, and enjoyment of outdoor and wilderness areas.

1001 SW Klickitat Way, Suite 201, Seattle, WA 98134

Published simultaneously in Great Britain by Cordee, 3a DeMontfort Street, Leicester, England, LE1 7HD

Manufactured in the United States of America

Project Editor: Christine Ummel Hosler
Copy Editor: Sherri Schultz
Cover, Book Design, and Layout: Mayumi Thompson
Cartographer and Illustrator: Moore Creative Designs

All photographs by the authors unless otherwise noted.

Cover photograph: *The dramatic escarpment at Bryce results from headward erosion of numerous streams draining a fault scarp along the east edge of Utah's Paunsaugunt Plateau.* Lucy Chronic photo.
Frontispiece: *Undermined by erosion of soft rock layers near the canyon floor, parts of the thick, massive Navajo Sandstone break away along vertical joints, leaving the huge monolith of Zion's Great White Throne.* Ray Strauss photo.

Library of Congress Cataloging-in-Publication Data

Chronic, Halka.
  Pages of stone : geology of Grand Canyon and plateau country national parks and monuments.— 2nd ed. / Halka Chronic and Lucy Chronic.
      p. cm.
  Includes bibliographical references and index.
   ISBN 0-89886-680-4 (pbk.)
  1. Geology—West (U.S.) 2. Geology—Great Plains. 3. National parks and reserves—West (U.S.) 4. National parks and reserves—Great Plains. 5. Natural monuments—West (U.S.) 6. Natural monuments—Great Plains.
  I. Chronic, Lucy II. Title.
  QE79.C47 2004
  557.8—dc22
                          2003022135

CONTENTS

Prefaces   9

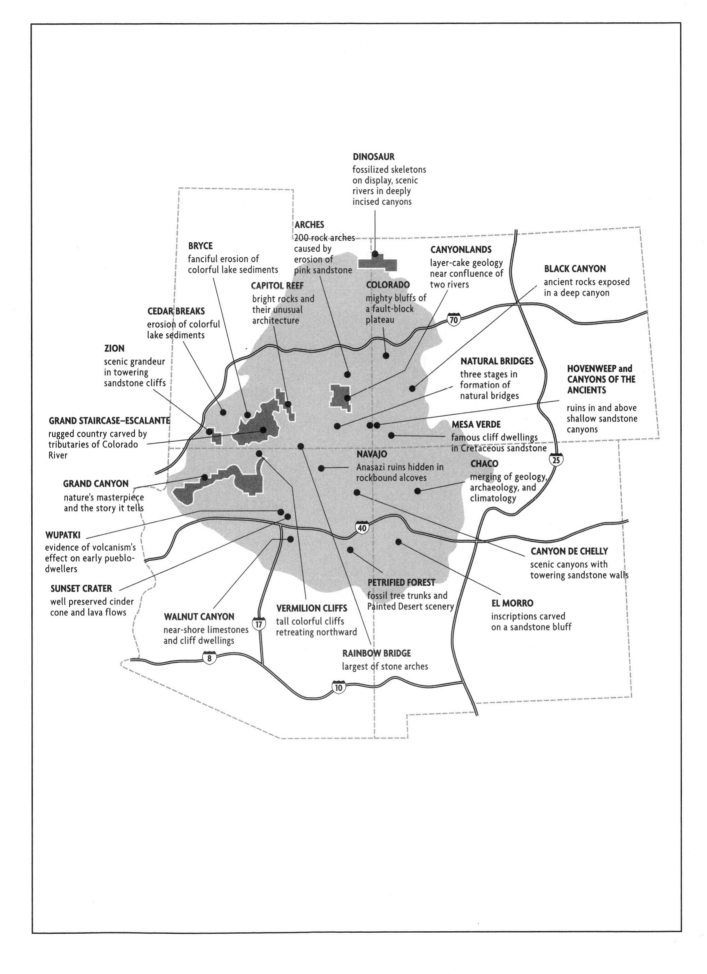

DINOSAUR
fossilized skeletons
on display, scenic
rivers in deeply
incised canyons

ARCHES
200 rock arches
caused by
erosion of
pink sandstone

CANYONLANDS
layer-cake geology
near confluence of
two rivers

BLACK CANYON
ancient rocks exposed
in a deep canyon

BRYCE
fanciful erosion of
colorful lake sediments

CAPITOL REEF
bright rocks and
their unusual
architecture

COLORADO
mighty bluffs of
a fault-block
plateau

CEDAR BREAKS
erosion of colorful
lake sediments

NATURAL BRIDGES
three stages in
formation of
natural bridges

**HOVENWEEP and
CANYONS OF THE
ANCIENTS**
ruins in and above
shallow sandstone
canyons

ZION
scenic grandeur
in towering
sandstone cliffs

GRAND STAIRCASE–ESCALANTE
rugged country carved by
tributaries of Colorado
River

MESA VERDE
famous cliff dwellings
in Cretaceous sandstone

NAVAJO
Anasazi ruins hidden in
rockbound alcoves

CHACO
merging of geology,
archaeology, and
climatology

GRAND CANYON
nature's masterpiece
and the story it tells

WUPATKI
evidence of volcanism's
effect on early pueblo-
dwellers

CANYON DE CHELLY
scenic canyons with
towering sandstone walls

SUNSET CRATER
well preserved cinder
cone and lava flows

PETRIFIED FOREST
fossil tree trunks and
Painted Desert scenery

EL MORRO
inscriptions carved
on a sandstone bluff

WALNUT CANYON
near-shore limestones
and cliff dwellings

VERMILION CLIFFS
tall colorful cliffs
retreating northward

RAINBOW BRIDGE
largest of stone arches

As shown abundantly in Canyonlands National Park, hard and soft rock strata erode into cliff-slope-cliff (or ledge-slope-ledge) sequences characteristic of the Colorado Plateau. Slopes of shale and mudstone undermine retreating cliffs of sandstone or limestone. Ray Strauss photo.

PREFACE TO THE SECOND EDITION

It is memories, I suppose, that shape a person's passion for place—one's own memories, but also others' memories. And Plateau country is, for me, a world filled with both. Now, superimposed on past memories are the impressions of this project, for it has been a joy to work with my mother, Halka Chronic—learning from her, teaching her, watching her. She has been and continues to be one of my greatest educators.

My mother and I want to thank all the additional people who have helped us with this edition. Discussions with Ivo Lucchitta, Ronald Blakey, and Stanley Beus clarified the geology of the Grand Staircase and the history of the Grand Canyon. Members and volunteers of the National Park Service responded to our many questions and supplied us with materials, and librarians at the Cline Library, Northern Arizona University, readily helped us with photos from its collections. Tad Nichols's and Ray Strauss's photographs continue, with additions, to grace these pages, and Jim Greslin and Felicie Williams supplied additional photos for this edition. Chris Hinze read and critiqued portions of the manuscript.

—*Lucy Chronic*

PREFACE TO THE FIRST EDITION

My earliest memory of the Plateau country is of a sunny summer's day in the late 1930s. I was driving with my mother from California to Massachusetts—quite an undertaking in those days—on the Lincoln Highway, which ran from San Francisco to the tip of Cape Cod. In southern Utah we diverged from the highway to gaze in delight at the sheer-walled magnificence of Zion and the breathtaking intricacy of Bryce Canyon's sculptured turrets, as lovely a landscape then as it is now.

I'd been in the Plateau country before, though, when my parents accompanied Dr. Byron Cummings—father of Southwestern archaeology—into Hopi and Navajo lands east of the Grand Canyon. A procession of cars—for the sake of science but not without the lure of adventure. Not much in the way of roads, I learned later. Turns were taken pulling and pushing each car through mud and sand. Our Studebaker brought up the tail end of the cavalcade. When the expedition bogged down, Mother knelt beside the gearshift and the hand brake to mix my powdered formula and warm it over a Sterno flame. I was six months old.

I slept cozily in a hammock slung between the car doors. My brother slept on the front seat, Mother on the back. Expedition tents sheltered flour and sugar and the precious finds of the archaeologists. My father (and, as he put it, God) slept out in the rain.

The expedition took in Hopi Indian snake dances at mesa-top Walpi. We two little ones stayed with an Indian nurse in Oraibi, our parents blissfully unaware of a measles epidemic there. In Kayenta (in those days pronounced kay-yen-TAY) we stayed with the Wetherills of trading-post fame. Dad and "Old John" Wetherill plunged into a friendship that lasted many decades. From there, my mother rode on horseback to Monument Valley (now a Navajo Tribal Park) with a guide named Harvey Adair—the same Harvey Adair who, when buying a neckerchief at the trading post, requested, "None o' yer gaudy colors. Jes' plain red 'n' yaller." Mother told later of the excruciating discomfort of three days in the saddle, and how she shifted a small pillow to ease the chafing, trying at the same time to absorb and memorize the beauty all around her.

We all went to Grand Canyon a year or so later—I've seen the photographs. We camped on the rim—the very rim. No paved roads there, either, and no organized campgrounds.

The road to Monument Valley acquired its pavement during the uranium boom of the 1950s and 1960s—but not before my own children had bounced over its dusty washboards and heard the story of their grandmother's horseback visit. And not before I became reacquainted with the Plateau country and, indeed, interested in geology through three summers at the Museum of Northern Arizona in Flagstaff.

And what a place to learn geology! Half the history of the Earth is here—from the dark, hard rocks of Grand Canyon's Inner Gorge, through the canyon's multilayered walls, up the "Grand Staircase" of the Vermilion Cliffs, White Cliffs, and Pink Cliffs in southern Utah, to the volcanic lavas of Sunset Crater, scarcely touched yet by erosion. Cliff-edged mesas and layer-cake plateaus, each band of rock younger than those below. Rocks alive with color: sandstone and siltstone and limestone. Rocks marked with curly imprints of ancient worm trails, dotted with fossil shells, creased with polygons of ancient mud cracks, festooned with ripple marks, transected by volcanic dikes. Seabed sediments high above the sea, dunes of an ancient Sahara, a fossil forest downed by an ancient cataclysm. Rocks worn by rain and snow and frost, deeply gouged by rivers, incised by seldom-flowing streams, etched by wind.

The Grand Canyon was "discovered" by one of Coronado's lieutenants as early as 1541. It was known much earlier by Native Americans, some of whom lived and farmed in its depths. One of the first Americans to explore this region was Lieutenant Joseph C. Ives, employed by the U.S. government to investigate the geology of the area. Ives was unimpressed:

Ours is the first, and doubtless the last, party of whites to visit this profitless locality. It seems intended by nature that the Colorado River, for the major part of its lonely and majestic course, shall be forever unvisited and undisturbed.

But the Grand Canyon was not dead. In 1869 John Wesley Powell, a one-armed veteran of the Civil War, led a daring expedition down the unexplored Green and Colorado Rivers and through the Grand Canyon. Powell firmly believed that a river carrying so much sand and silt and rock rubble would have long ago ground down any waterfalls. Fortunately he was right! The trip took longer than planned, for Powell, though self-educated, was a thorough scientist and documented every phase of the expedition with accurate measurements and vivid descriptions of rocks and river and ruins. Before he and his party emerged from the western end of the Grand Canyon, newspapers across the country brandished obituary notices (about which Powell later joked, "In my supposed death I had attained to a glory which I fear my continued life has not fully vindicated"). Like the Grand Canyon, Powell was very much alive. Partly on the basis of his Plateau country exploration, he went on to urge Congress to establish the U.S. Geological Survey, an agency whose first task was to further explore the West.

Most of the parks and monuments described in this book by their very nature emphasize geology. A few—Mesa Verde, Chaco, Navajo, El Morro—place emphasis on human history and prehistory, but even in them the rocks stand out boldly, colorfully, often dramatically, as part of the even earlier history of the Colorado Plateau.

To learn about their geology, visit these parklands—both parks and monuments—for all they are worth. See introductory films and slide shows offered by the visitor centers. There you can also find topographic and, for some areas, geologic maps, as well as other literature discussing geology. For finding your way around, the small maps distributed at entrance stations by the National Park Service are quite sufficient. With few exceptions, places mentioned in this book appear on these maps.

You'll learn also, as Powell did, by experiencing the geology for yourself, exploring parks and monuments by car and on foot, getting out and seeing with your own eyes, touching with your own fingers. (Note: Collecting rocks, minerals, fossils, and plant and animal material is not permitted in national parks and monuments.) Roads and trails (some with guide leaflets) visit specific geologic features described in this volume. Allow yourself time for exploring these features, as well as for discovering others on your own.

Geology is a logical science, and by pausing long enough to study landforms and rocks, by looking at them carefully, by thinking about what you are seeing, you may be able to interpret some of them for yourself. Use the present as the key to the past. Streams and rivers of the past washed rocks and sand and silt from mountains and highlands, and deposited them on floodplains and deltas, just as modern rivers and streams do. Sand and clay and limy mud deposited in the past have, with time and the weight of other rocks above them, hardened into sandstone and mudstone and limestone, just as sediments deposited today will in the end, if they are not washed away, become rock. Sands of ancient deserts have counterparts in today's Sahara, or even in dunes here and there in our own Southwest. Winds and flowing waters of the past rippled mud and sand, and raindrops pitted soft surfaces with tiny craters, just as their modern counterparts do. Shellfish of the past, preserved as fossils in rock, have modern descendants. Other animals left other exciting glimpses into their lives—footprints in the sand, now preserved in rock.

Although this book is designed for readers without formal geologic education, I hope it will prove useful to students and professionals as well. In Part 1, geologic terms are printed in bold type where first used and, unless their meaning is clear, defined. Most of the terms are defined again in the glossary in the back of the book. Additional reading, suggested at the end of each section, emphasizes articles, books, and maps that can be understood with no more geologic background than that presented in Part 1.

The information presented here is drawn from published geologic literature, from discussions and correspondence with individual geologists who have worked in and near park areas, and from personal observation. I sincerely thank the many colleagues whose work I leaned on, as well as those who helped me more directly. I also express my gratitude to members of and volunteers in the National Park Service, who made my visits to parks and monuments more interesting and exciting, and who in addition reviewed portions of the manuscript. My special thanks go to Richard Hereford of the U.S. Geological Survey, and to Faye and Arthur Frost, who read and critiqued the entire manuscript and offered many helpful suggestions; to Tad Nichols and Ray Strauss, who contributed many of the photographs; and to Luiz Schleiniger, who produced good prints from my often mediocre negatives. Felicie Williams provided one of the color photographs, and Emily Silver, also my daughter, prepared many of the figures and diagrams.

Writing this book has been a rich and rewarding experience. Not least of the pleasures has been the delightful excuse to hike again the trails of Grand Canyon, to raft the rapids of its great river, to gaze once more on the breathtaking beauty of Bryce, to explore the hidden clefts and chasms of Zion, to prowl the back roads of Canyonlands, and, perhaps most deeply memorable, to stand alone in silent awe, hearing the whispers of the Ancient Ones at the cliff dwellings of Navajo National Monument.

—*Halka Chronic*

A NOTE ABOUT SAFETY

Safety is an important concern in all outdoor activities. No guidebook can alert you to every hazard or anticipate the limitations of every reader. Therefore, the descriptions of roads, trails, and natural features in this book are not representations that a particular place or excursion will be safe for your party. When you visit any of the places described in this book, you assume responsibility for your own safety. Under normal conditions, such excursions require the usual attention to traffic, road and trail conditions, weather, terrain, the capabilities of your party, and other factors. Keeping informed on current conditions and exercising common sense are the keys to a safe, enjoyable outing.

—*The Mountaineers Books*

PART 1

OF EARTH AND TIME

I. THE COLORADO PLATEAU

The Plateau country, a land famous for its tranquil beauty, is geologically a raft in a stormy sea. A broad, high tableland, it has steadfastly resisted forces that bent and crumpled the surrounding country and created waves of mountains to the north, east, south, and west. Bypassed by mountain-building forces, its rocks, lifted to elevations as great as 10,000 feet (3000 meters), still lie in the horizontal positions in which they formed.

Properly called the Colorado Plateau, honoring the great river that has carved dramatic canyons through it, this region centers around the Arizona–Utah border but reaches into western Colorado and northwestern New Mexico as well. Throughout this book, "The Plateau," spelled with a capital P, refers to the Colorado Plateau as a whole. The Plateau is made up of many smaller **plateaus**—Shivwits, Coconino, Kaibab, Paunsaugunt, Markagunt, and others—each a few

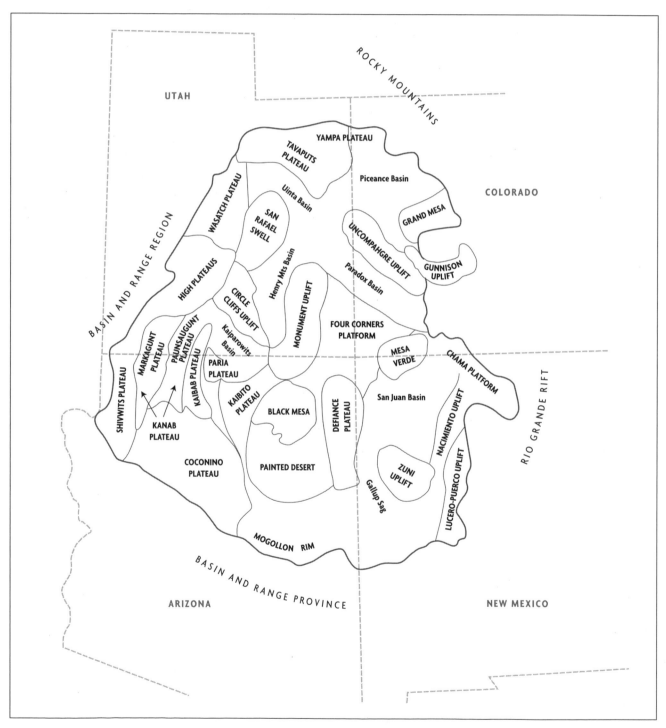

The Colorado Plateau covers a four-state area and consists of many smaller individual plateaus and a few intervening basins.

In the Painted Desert, the very nature of the rocks prevents plant growth and brings about badland erosion. Here, blocks of petrified wood weather out and roll down into the gully. Ray Strauss photo.

hundred feet higher or lower than its neighbors. These smaller plateaus, like the logs of a loosely bound raft, have shifted upward and downward along major breaks in the Earth's crust. They vary in character with elevation, with types of rocks exposed at the surface, and with amounts of rainfall or snowfall—usually functions of elevation. Surfaces of these smaller plateaus range from 4500 to 10,000 feet (1500 to 3000 meters) above sea level. Some are separated by river valleys or wide basins, others by lines of cliffs—Pink Cliffs, White Cliffs, Vermilion Cliffs, Straight Cliffs, Circle Cliffs, and so on. All but the highest of the plateaus are arid or semiarid, deprived of water by the great barrier of California's Sierra Nevada, which during much of the year intercepts moisture from the Pacific. In only two seasons does this region receive appreciable moisture: in winter, when storms come in from the northwest, and in summer, when monsoon-type winds sweep in from the south, avoiding the mountain barrier.

Long dry periods between these seasons limit plant growth. In the lowest, hottest parts of the Plateau, sage, saltbush, and other low shrubs dominate, and desert grasses turn green with seasonal rains. Pinyon-juniper woodlands are found at intermediate elevations. And pine, spruce, fir, and aspen forest the highest plateaus.

Characteristically, deserts develop in areas of low rainfall. But on the Plateau another factor comes into play: the chemistry of the rocks. Deserts here may be due to gypsum or salt in the rocks, or to **clays** derived from **volcanic ash.** Some of these clays swell when they are wet and become soft and crusty when they dry, making plant growth difficult. And without plants to help hold down the soil, unconstrained **erosion** creates **badlands** like those of Arizona's Painted Desert (see sidebar on page 132).

On other parts of the Plateau, too, especially in canyons and along cliffs that edge different plateau levels, rocks are clearly exposed for long distances. And this means color, for the rocks of this region are gaily tinted in shades of red, pink, purple, green, and yellow. It also means a paradise for geologists.

The rocks of the Colorado Plateau may in places be darkened with **desert varnish,** a thin, shiny, blue-black polish of iron and manganese oxides. Desert varnish develops slowly, over many centuries, either from minerals gradually leached from the rock or from soluble material carried in dust and spread thinly, over and over again, by occasional rain. **Lichens,** too, color rock surfaces with green and orange blotches, or streak them top to bottom with long black bands that indicate seepage lines where rock remains wetter for longer periods after rain (see sidebar on page 66).

Here as elsewhere, erosion works on exposed rock surfaces. In this land of sparse rain and snow,

Long streaks of black lichens and brown or blue-black desert varnish stain many Plateau country cliffs. These are in Canyon de Chelly National Monument. Halka Chronic photo.

weathering, the decomposition and disintegration of rock, proceeds in some cases more rapidly than you might think, providing abundant cargoes of **sand** and mud for ephemeral streams. As in all desert regions, however, soils are thin and vegetation is for the most part scanty. **Landforms** on the Plateau therefore look different from those found in moist climates. Hardness and softness of different rock layers markedly influence erosion, and one characteristic of the Plateau is the alternation of hard and soft rock layers. Cliffs and ledges of resistant **limestone** or **sandstone** alternate with slopes and **benches** of less resistant **shale** or **mudstone.** On both large and small scales, such **differential erosion** adds to the variety of Plateau landscapes.

Outcrop patterns, in a land of horizontal rock layers, follow the contours. Cliffs slowly retreat as they

Where wind blows away smaller rock particles, pebbles concentrate as desert pavement. Halka Chronic photo.

are undermined by erosion of weak layers below. As rocks break away from precipices, individually or in massive rockslides, they lie for many centuries just where they have fallen. Or, tumbling to a canyon bottom, they serve as tools for streams coming to life during sudden thunderstorms, helping the rushing waters gouge steep, angular canyons by hammering rock against rock and rasping rock with sand.

Canyons on the Plateau are steep-walled and nearly barren of vegetation. Without the moisture that lubricates landslides or slows soil-creep of wetter climates, canyon profiles maintain the cliff/slope/cliff alternation described above. At higher elevations, the freezing and thawing of water held in cracks and crevices helps to break down the rock.

In limestone, seeping water trickling through tiny cracks and crevices gradually dissolves part of the rock, slowly enlarging its crevices and shaping corridors and twisting passageways. **Sinks** and **solution valleys** are caused by the collapse of such passages.

Wind, too, plays a strong role in erosion here. Sweeping across wide plateau surfaces, picking up sand and small pebbles, it hammers at cliffs and ledges, emphasizing in strong relief the strengths and weaknesses of the rock. Whirling in summer dust devils, wind with its airborne tools quickly frosts rock surfaces (and car windshields) with tiny pits. On broad slopes, wind and rain create **desert pavement,** a thin armor of closely spaced pebbles left behind when finer material—sand and silt—has blown or washed away.

II. PLATEAU ARCHITECTURE

Many individual plateaus are outlined by **faults** and **folds. Faults** are breaks in the Earth's crust along which movement has taken place. Most of the faults

Honeycomb weathering forms as abrasive, sand-laden winds bombard sandstone or siltstone already pitted by solution of salt inherent in the rock.
Lucy Chronic photo.

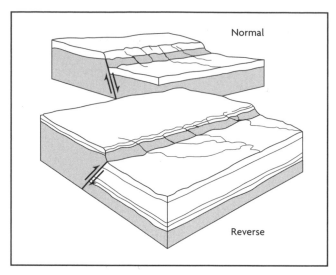

Most of the faults of the Plateau Country are nearly vertical normal and reverse faults.

that separate individual plateaus trend north and south, some in zigzag fashion. Most are steeply inclined **normal** or **reverse faults;** their relative up-and-down motion controls the heights of adjacent plateaus. Because of cliff retreat in this semiarid country, the actual faults may be some distance out from the base of the cliffs that edge those plateaus.

Faults come in many shapes and sizes. In addition to large faults that separate individual plateaus, thousands of much smaller faults are present in all parts of the plateau area. The smallest may represent movements of less than a fraction of an inch. Cracks or fractures that show no movement at all—and they are abundant in nearly all rock—are called **joints.**

Folds, on the other hand, are bends rather than breaks. Some plateaus are separated by **monoclines,** the simplest of folds, with one side raised higher than the other. Exposed rocks in the depths of the Grand Canyon show us that most monoclines are surface expressions

Faults frequently consist of clusters of small offsets, as seen here between arrows that denote direction of fault movement. Note that evidence of faulting disappears in the dark, soft mudstone at the left. Halka Chronic photo.

In many parts of the Plateau country, massive sandstone fractures along vertical joints, spilling large angular fragments onto talus slopes below. These are in Capitol Reef National Park. Ray Strauss photo.

of faults far below, as shown in the illustration (see also sidebar on page 112).

Like faults, folds come in many sizes and shapes. On the Plateau we find, in addition to monoclines, gentle downward warps or **synclines,** equally gentle upwarps or **anticlines,** and more or less circular, more or less equidimensional downwarps and upwarps called **basins** and **domes.**

The Plateau, with its well-exposed, straightforward geology, is a geologic textbook—the "pages of stone" that give this book its name. Many geologic features—the fault/fold relationship described above, for instance—can be studied in detail without having to hack away thick underbrush or dig through soil cover. And deep canyons, with their tributaries, give us three-dimensional views. Barren rocks sing out with features that tell us where and how they formed—in seas or on land, on beaches or in lagoons, on deltas or deserts. They even tell us which way rivers flowed and in what direction wind blew in the remote geologic past. What's more, we can know for sure that in an area so undisturbed, the oldest rock layers are on the bottom, the youngest on top—one of the basic tenets of geology.

Clearly, geologic architecture gives us our scenery here. The plateaus have risen and fallen; the forces of weathering and erosion—water, wind, and frost—have shaped cliffs and spires, worn little gullies and great canyons, etched and sharpened hard rocks, washed away softer and less resistant ones. To understand the basic causes of these changes we need to pause and look at the Earth as a whole, to see what is going on beneath our planet's crust that can influence developments on its surface.

III. THE EARTH'S INTERIOR

Hidden from view, the interior of the Earth long remained a mystery. Very gradually, geologists have found clues to its makeup in earthquake waves and in determinations of its total mass as indicated by its gravity.

From these clues, we have learned that the Earth's **core** is a sphere of almost pure nickel and iron, its center solid but its outer part semiliquid, capable of some degree of plastic movement, as is, for instance, malleable red-hot iron on a blacksmith's anvil. The core is not necessarily smooth-surfaced; research shows that it boasts higher mountains and deeper valleys than the Earth's surface.

Outside the core is another thick layer, the **mantle.** It is composed of **basalt,** the same type of black or dark gray material, rich in iron and magnesium, that occasionally spills out on the Earth's surface as dark gray or black **lava.** As was the case for the core, the inner part of the mantle is solid, the outermost part liquid or semiliquid—a seething, slow-boiling, red-hot layer. The mantle is the Earth's stove, heated by the decay of radioactive minerals concentrated there. Con-

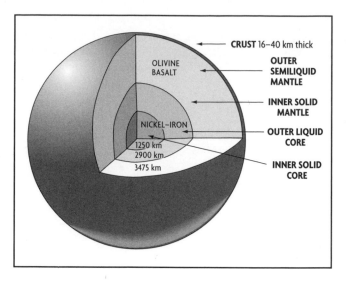

The Earth is composed of several distinct compositional layers, some liquid and some solid.

vection currents develop in the mantle in the form of huge cells that surge upward in gigantic boils, roll over, and then plunge downward again.

Powered by its own hot stove, the Earth's churning mantle exerts a strong influence on the outermost shell of the Earth, the **crust.** The crust is made up of what we think of as solid rock. On a worldwide scale, there are two major types of crust. The first type, **oceanic crust,** is dark gray or black and heavy with iron and magnesium, very like the mantle in composition. It occurs, as its name suggests, beneath ocean basins. Oceanic crust is quite thin, something on the order of a mere 3 miles (5 kilometers).

The other type of crust, more familiar to us because we live on it, is **continental crust.** Lighter than oceanic crust in both color and weight, it is composed of minerals low in magnesium and iron. It occurs (as you've no doubt guessed) on continents, extending out to the limits of the continental shelves. Continental crust is relatively thick, about 20 to 25 miles (30 to 40 kilometers).

IV. CONTINENTS ON THE MOVE

The idea that continents move, or drift, was born early in the twentieth century—a result of recognition of the match in profiles of the continents bordering the southern Atlantic Ocean: Africa and South America look like two adjacent pieces of one jigsaw puzzle. And if you juggle other pieces a little bit, particularly if you include their continental shelves, Europe and North America fit together, too. What's more, if you use geologic maps, which show the nature of rocks at the surface, the fit is even better: Adjoining coastlines show many of the same rock types and ages. So some researchers supported the idea that the continents had once been connected, but had broken apart and drifted away from each other.

The catch was that nobody could come up with a plausible theory to explain how parts of the thin, fragile crust could sail off around a globe of solid, or nearly solid, mantle.

In the 1960s, however, geologists developed new views, born, like most scientific views, of the old. The **Theory of Plate Tectonics** answered so many previously perplexing questions that it took the geologic world by storm. Geologists everywhere seized upon it to explain their particular findings in their particular parts of the world.

This theory once again sets the continents adrift. But it portrays the crust as coupled with the uppermost part of the mantle to form a stiff layer, the **lithosphere** ("rock sphere"), a layer about 40 miles (60 kilometers) thick under the ocean and about 60 miles (90 kilometers) thick under the continents. Relative to the size of the Earth, that's still pretty thin—hardly more than a film. But like the film of congealed fat on a bowl of chicken broth, it can be rumpled and moved about, or even broken, by stirrings in the mantle.

The lithosphere, and therefore the crust, is made up of a dozen large **plates** fitted together like the plates of a turtle's shell, with a bevy of smaller plates squeezed in here and there among the large ones. Between the plates are boundaries that in some cases are the submerged volcanic mountain chains of **mid-ocean ridges,** and in other cases are deep-sea **trenches** or magnificent high-rise ranges like the Alps and the Himalaya. All these boundary areas are beset with frequent earthquakes and periodic volcanic activity, to the extent that maps showing the location of earthquakes and volcanoes distinctly outline the tectonic plates.

The theory also states that new crust is constantly coming into being along the mid-ocean ridges that thread their way down the Atlantic, through the Indian Ocean, and across the Pacific. The ridges bear along their summits narrow volcanic fissures where molten rock or **magma** boils up from the mantle and hardens into new oceanic crust. Created in two narrow bands, the new crust slides apart as still newer lava erupts along the fissures. This continuous replenishment makes of the ocean's floor two broad conveyor belts moving slowly and in opposite directions away from each mid-ocean ridge. Movement of the conveyor belts comes about because plates on each side of the mid-ocean ridges are dragged apart by slow-moving but powerful convection currents stirring the semimolten part of the mantle. The whole process, which we now call **seafloor spreading,** is a vital part of the Theory of Plate Tectonics.

To compensate for new crust born along mid-ocean ridges, old crust is destroyed near the continents. Where a continental plate collides with an oceanic plate, the oceanic plate, made of heavier material, goes under. It is overridden by the lighter (though thicker) continental plate, and is then drawn down, **subducted,** to depths where it remelts and once more becomes part of the mantle. (Rock melts at temperatures of 2000 to 2200 degrees Fahrenheit, or 1100 to 1200 degrees Celsius.) Where both colliding plates are continental, mountains are pushed up, foreshortening the crust. This has happened, for instance, where India, once a small continent in itself, bumped into Asia with enough force to push and crumple and thrust up the Himalaya.

In the oceanic plate–continental plate type of collision, volcanism is likely to occur near the line where subduction takes place, with magma rising from two sources: remelted oceanic crust, which gives rise to dark, iron-rich magma; and portions of the leading edge of the continental plate, drawn down and remelted along with the oceanic crust. The more buoyant continental

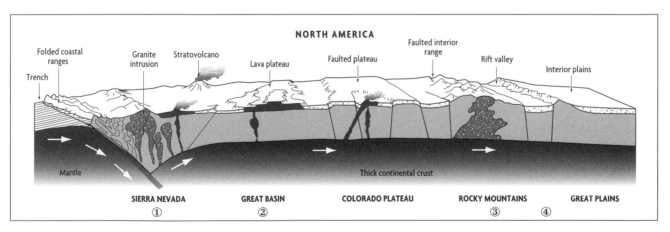

1. *As oceanic crust is subducted, melting of continental crust along the subduction zone creates granitic magma, which may cool slowly, below the surface, in batholiths that are later bared by erosion. The magma may erupt explosively to form stratovolcanoes.*
2. *Flood basalts, exceptionally fluid in nature, rise above "hot spots" below continental crust.*
3. *Interior ranges push upward in response to compression along the distant continental margin.*
4. *Rift valleys are tensional features where the crust drops as neighboring areas are pulled apart.*

material, once melted, may push upward through the crust near the edge of the continent. If such magma rises only partway to the surface, and cools while it is still at considerable depth, it hardens—crystallizes—slowly into great masses of granite. Where it breaks through to the surface, volcanoes form, with light-colored lava unlike the dark lava derived from oceanic crust or directly from the mantle.

The North American Plate, like most other large plates, is partly oceanic (its eastern half, under the Atlantic Ocean) and partly continental (its western half). For the last 60 or 70 million years, as new crust formed along the Mid-Atlantic Ridge and the Atlantic's seafloor spread, the North American continent drifted westward and southwestward at a rate of about 2 inches (4 centimeters) per year. Long ago its westward drift put it on a collision course with the Farallon (or East Pacific) Plate

and some smaller plates clustered along it.

Because the western part of the North American Plate is composed of continental crust, lighter in weight than the oceanic crust of the Farallon Plate, the North American Plate overrode the Farallon Plate, forcing it downward to remelt in the mantle. As this happened, the advancing edge of the continent collected a number of large islands or microcontinents, adding them to its western margin, so the continent gradually became wider in a westward direction. Eventually the North American Plate overrode almost all of the Farallon Plate, even covering large portions of the East Pacific Rise, the Pacific's mid-ocean ridge. When the North American Plate came in contact with the plate west of the mid-ocean ridge, the northwestward-moving Pacific Plate, faulting such as that along the San Andreas Fault replaced subduction.

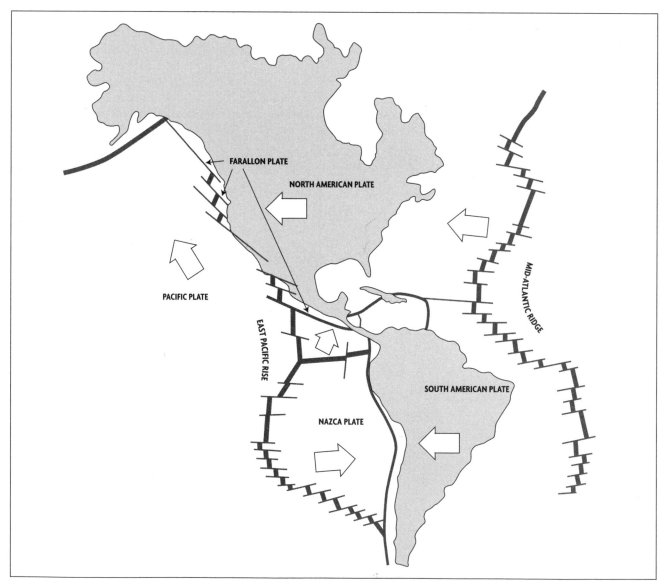

With formation of new crust along the Mid-Atlantic Ridge, the North American Plate moves westward (arrows) across segments of the Farallon Plate, now mostly hidden under western North America. Mid-ocean ridges are offset by numerous transform faults.

But just where does the Colorado Plateau fit into this picture? The Plateau seems to have resisted the forces involved in mountain building and to have maintained its integrity right through the westward drift and the collision of the North American and Farallon Plates, as well as through several earlier movements in its history. On the Plateau, rock layers ranging in age from 600 million to a mere 6 million years are still in the horizontal position in which they were deposited. Seismic studies, which measure the thickness of the crust by the velocity of earthquake waves that pass through it, show that the crust is thicker under the Plateau than under the Basin and Range region to the north, west, and south, which, along with rock strength, may contribute to its raftlike integrity and to its longevity.

The **uplift** that raised the Plateau, with all its segments, to its present altitude seems to have resulted from general uplift of the western part of the continent, perhaps due in turn to crustal thickening, or to the upwelling of heat cells in the mantle as the continent drifted westward over the East Pacific Rise.

The Plateau as a whole seems to be rotating horizontally, like a wheel on its side, dragged in a clockwise direction by northward movement of the Pacific Plate against the edge of the North American Plate. The circular movement has torn the continent along the Rio Grande Rift of New Mexico, moving the Plateau away from the rest of the continent east of the rift. By and large, the Plateau seems to have been subjected to the same strains and stresses that governed faulting and collapse in the Basin and Range region to the west, though the thick, strong crust of the Colorado Plateau responded differently to these stresses than did the Basin and Range region.

V. ROCKS AND MINERALS

Geologists recognize three main classes of rocks, all of them represented on the Colorado Plateau. Since sedimentary rocks are by far the most common type in this area, we'll look at them first.

Scalloped by wind, these ripple marks grace a sandstone slab in Capitol Reef National Park.
Halka Chronic photo.

Long, sweeping cross-bedding tells of windblown sand that accumulated as sand dunes. Crumpled layering at the top of this sandstone may indicate earthquake activity while the sand was still soft.
Halka Chronic photo.

- **Sedimentary rocks** form from the broken fragments or dissolved minerals of other rocks, transported and deposited by water, wind, or ice. Nearly always, sedimentary rocks are layered or **stratified,** which makes them easy to recognize. (**Lava flows** and falls of volcanic ash may be stratified, too, though they are classed as volcanic rocks.) Nearly always, the stratification was originally horizontal or nearly so.

  Sedimentary rocks are classified by grain size and composition. They range from very fine-grained **claystone** to coarse **conglomerate** made of pebbles and **cobbles** imbedded in a finer matrix. They also include **limestone** and **dolomite** made not of rock fragments but of calcium carbonate and magnesium carbonate precipitated chemically or derived from literally billions of plant and animal shells. Sedimentary rocks frequently contain **fossils,** remains or traces of animals and plants that lived and died when the rock was forming. Sandstone and siltstone deposited by flowing water or wind may be **cross-bedded,** with fine **laminae** slanting at an angle to the stratification. Or they may preserve **ripple marks, mud cracks,** and **raindrop imprints** that formed on their surfaces as they were deposited, and that are important clues to their mode of origin.

  Sedimentary rocks are also classified by location of deposition. They are **marine** if they are deposited in the sea, and **continental** if they are deposited on land by rivers, lakes, glaciers, or wind.

- **Igneous rocks** originate from molten rock material known as magma, which rises from as much as 200 miles (300 kilometers) below the surface. Igneous

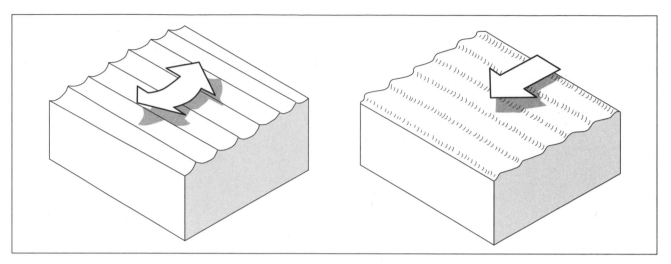

Ripple marks are formed by moving water or wind. Current-formed ripples (right) differ in shape from those formed by oscillating water (left).

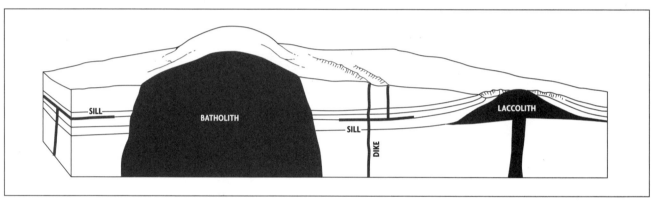

SILL BATHOLITH SILL DIKE LACCOLITH

Molten rock—magma—that cools below the surface becomes intrusive rock. Intrusions come in many shapes and sizes, and may later be bared by erosion, as shown here.

rocks are further divided into two groups: those that cool and harden very slowly without reaching the surface, called **intrusive igneous rocks,** and those that harden more rapidly at the surface, called **extrusive igneous rocks** or just **volcanic rocks.**

Long, slow cooling promotes crystal growth, so intrusive igneous rocks are generally grainy, with easily distinguished, tightly packed mineral crystals. Though they vary in composition, these grainy rocks can all be called granite. Granite is relatively rare in the Plateau country but can be found in the depths of the Grand Canyon, in Black Canyon of the Gunnison National Park, and in Colorado National Monument. It also occurs in lenslike **laccoliths** that form many small mountains in the Plateau region.

Thin sheets of intrusive igneous rock may also be sandwiched between layers of sedimentary rock as **sills,** which parallel the stratification of the sedimentary rock, and as **dikes,** which cut across rock layers. Both sills and dikes are crack fillers, either injected with enough pressure to force the older rocks apart, or flowing into open joints or faults. Since sills and dikes, as well as small laccoliths, cool

more rapidly than the larger masses associated with major mountain ranges, their crystal grains are smaller.

Volcanic rocks are fairly common on the Colorado Plateau. Basalt, appearing in lava flows and as volcanic cinders, is dark in color and quite fluid or foamy when it erupts. Some basalt lava flows, having followed stream valleys, are long and narrow. Others spread out over sizable parts of the Plateau.

Lighter in color than basalt, **silicic** volcanic rocks are much less fluid and contain more gas when they erupt. They form short, thick lava flows, mound into **lava domes,** or plug up a volcano's **vent** until they are exploded away. Alternating with layers of volcanic ash resulting from explosive eruptions, silicic lava flows may pile up into sizable **composite volcanoes** or **stratovolcanoes,** of which there are several on the Colorado Plateau.

• **Metamorphic rocks** form when preexisting sedimentary or igneous rocks are subjected to intense pressure and heat, usually stemming from deep burial or from folding and mountain building due

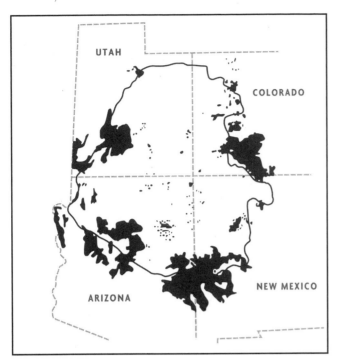

Volcanic rocks—lava flows, cinder cones, and stratovol-canoes—cover large areas near the Plateau margins. Smaller dots are eroded centers of older volcanoes.

to tectonic activity. Their grains may simply fuse together, in which case their origins can still be recognized. Or they may almost melt into new magma or recrystallize completely, so that it's quite difficult to deduce their original nature. In some, for instance, swirly dark and light banding suggests that they once were stratified. Whether the original layers were sandstone, siltstone, lava, or volcanic ash can often be determined by chemical analysis.

Two common types of metamorphic rock are **gneiss,** a coarse-grained, usually banded rock much like granite, and **schist,** a finer-grained rock that splits along irregular parallel planes because of its large component of the platy mineral **mica.**

Very old metamorphic rocks—both gneiss and schist—are found in Grand Canyon National Park, Colorado National Monument, and Black Canyon of the Gunnison National Park. They almost certainly underlie, along with granite, the rest of the Colorado Plateau.

All rocks are made of **minerals,** naturally occurring substances that have definite chemical makeups and often definite and characteristic ways of crystallizing or of breaking. By identifying one or two common

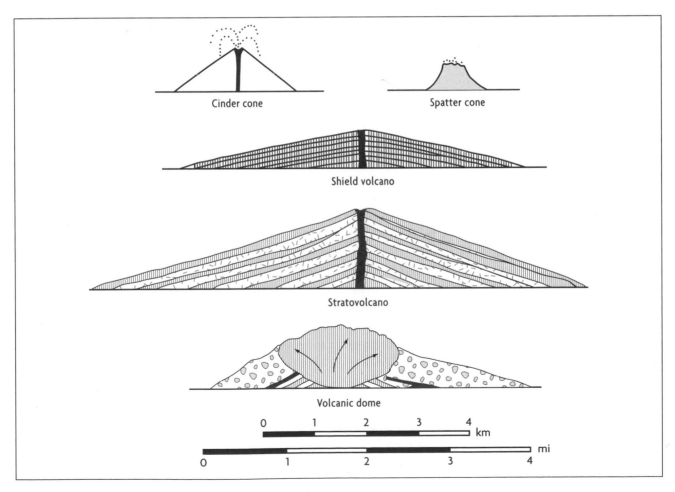

Volcanoes of several types occur on the Colorado Plateau.

minerals in each kind of rock, geologists refine rock descriptions, as for instance when they speak of **biotite schist** or **quartz sandstone.**

Even without a course in geology, you probably can already recognize quite a number of minerals: quartz, mica, native gold, gemstone minerals like ruby and diamond, semiprecious minerals like turquoise and agate. **Gypsum** and ordinary table salt are familiar minerals, too. **Quartz** and the light pink or gray members of the **feldspar** group are by far the most common of the rock-forming minerals. **Calcite** is the chief component of limestone and marble. Not many collector-quality minerals occur on the Plateau, and collecting of any kind is prohibited in national parks and monuments anyway, so minerals won't be emphasized in this book.

The almost gaudy rocks of the Plateau country owe their color variations to their mineral content. Tiny amounts of iron oxide—the minerals **hematite** (reddish brown) and **limonite** (mustard yellow), the same substances as red and yellow rust—give the rocks varying hues of pink and pale buff. Larger amounts of these same minerals produce deeper reds and brighter yellows. Many of these hues come about when **groundwater** percolates through rock that contains fine particles of black mica, **hornblende,** or other nonoxide iron minerals. When unoxidized, as when rocks form under anoxic (no oxygen) conditions, iron imparts a gray-green or purplish tone.

VI. GEOLOGIC DATING

Geologic time goes back to the formation of the Earth about 4.6 billion years ago—a pretty sizable chunk of time. But the rocks of the Earth's crust can be studied back to only about 3.8 billion years ago, when the oldest rocks now known were formed. The oldest rocks of the Colorado Plateau are the metamorphic and igneous rocks exposed in Colorado National Monument and in the depths of the Grand Canyon and the Black Canyon of the Gunnison; they are about 2 billion years old. Since they formed, continents have more than once drifted together and broken apart. Mountains have formed and worn away and formed again, only to wear away once more. Life has burst forth, first in the seas and then on land, slowly evolving into the plants and animals we know today.

But how do we know this? How can we recognize such immensities of time? How can we "date" rocks? Obviously, days, weeks, and years don't mean much against the immensity of geologic time, in which we deal with millions and billions of years. But we would like to know how many millions or, in some cases, thousands of years ago individual rock units formed.

Early dating methods relied on two things: the position of rocks relative to each other (we've already seen that in undisturbed sequences the oldest rocks are at the bottom, the youngest on top), and fossils,

relics or traces of life preserved in rocks. Dating by relative position or by fossils doesn't tell us exactly how long ago a given rock unit formed, but it does tell us which rock is older or which is younger—as long as the rocks are in approximately their original orientation. These techniques, however, work only for sedimentary and volcanic rocks, which always form from bottom to top. Intrusive igneous rocks tend to push up from below.

The use of fossils to date rock units came at a time when the evolution of plant and animal life had not yet been demonstrated. With Charles Darwin and a new understanding of evolution, earlier findings that certain recognizable fossils could be found in certain recognizable rock layers took on a new, two-way meaning: Successive fossil-bearing layers illustrate evolution, and evolution can be used to define the ordered sequence of rock layers on a worldwide basis.

With these dating techniques, haphazard though

ERA	PERIOD	EPOCH	AGE IN YEARS
CENOZOIC Age of Mammals	QUATERNARY Q	HOLOCENE Q	— 10,000 —
		PLEISTOCENE Q	
			— 1.8 million —
	TERTIARY T	PLIOCENE Tp	— 5.3 million —
		MIOCENE Tm	
			— 23.8 million —
		OLIGOCENE To	— 33.7 million —
		EOCENE Te	
			— 54.8 million —
		PALEOCENE Tp	
			— 65 million —
MESOZOIC Age of Reptiles	CRETACEOUS	K	— 144 million —
	JURASSIC	J	
			— 206 million —
	TRIASSIC	Ⱦ	— 248 million —
PALEOZOIC Age of Fishes	PERMIAN	Pm	
			— 290 million —
	PENNSYLVANIAN	P or ℙ	
			— 323 million —
	MISSISSIPPIAN	M	
			— 354 million —
	DEVONIAN	D	
			— 417 million —
	SILURIAN	S	
			— 443 million —
	ORDOVICIAN	O	
			— 490 million —
	CAMBRIAN	€	
			— 543 million —
PRECAMBRIAN P€	ORIGIN OF LIFE		3.5 billion
	ORIGIN OF EARTH		4.6 billion

Geologic time chart

they were at first, geologists put together a worldwide geologic calendar showing the relative ages of rocks, with names for months, weeks, and days of geologic time. In this calendar, the largest units, geologic months, are called **eras.** Eras are divided into **periods.** For detailed work, periods are divided into **epochs.** Names for major time units are shown in the chart on the opposite page. Epoch names, which vary from continent to continent, are given here for American Tertiary and Quaternary Periods only.

Geologists have now learned how to determine the absolute age of rocks with some degree of accuracy. This they do by measuring the decay of radioactive minerals such as uranium, carbon-14, or radioactive potassium-40, and their abundance relative to their own decay products. This technique is known as **radiometric dating,** and it gives us reasonably accurate dates of origin, in years, for tested rocks and minerals. By combining radiometric dating with the older calendar of geologic time, we now have a more accurate calendar on which to base our knowledge of the history of the Earth.

Another technique, **paleomagnetic dating,** relates natural rock magnetism to the pieced-together history of reversals in the Earth's magnetism, when the north and south magnetic poles switched their positive and negative polarity. The pattern that develops from switches in polarity (which are not at all regular with time, but are determined by studies of seafloor basalts in which iron minerals, oriented by the Earth's magnetism, were "frozen" into the rock) can also be added to the geologic calendar.

One more type of dating, applicable to quite young, quite recent deposits, relies on variations in the widths of tree rings. Used a great deal by archaeologists to date prehistoric ruins and artifacts, it is equally useful for dating prehistoric volcanic eruptions, floods, and other events that preserved trees or fragments of wood.

Except for very recent events, instead of using B.C. or A.D., geologists denote time as B.P.—"before present"— or as so many years "ago." Their figures are by their very nature approximate. A geologic event that happened 100 million years ago will not be significantly older next year.

Because of the way the geologic calendar developed, boundaries between its time divisions reflect notable changes in the history of the Earth and its inhabitants. The era divisions were meant to indicate stages in the development of life, from "early life" (Paleozoic) to "middle life" (Mesozoic) to "recent life" (Cenozoic). These three eras are also popularly called the Age of Fishes, the Age of Reptiles, and the Age of Mammals, named for their dominant animal types. When the Precambrian was first defined, life was thought not to have existed before Paleozoic time. We know now that both plants and animals did exist then, in highly complicated forms. The ancestors of all our major groups, as well as some groups without modern descendants, had already evolved, but did not secrete the hard shells and skeletons likely to be preserved as fossils (see sidebar on page 101).

Although the rock record is more or less continuous, with gaps in the record in one area filled in

MASS EXTINCTIONS

Extinctions, particularly the demise of the dinosaurs at the end of the Mesozoic Era, have puzzled geologists and paleontologists for decades. Other significant and widespread extinctions mark the end of Precambrian time and the Paleozoic-Mesozoic boundary, where differences in fossil animals and plants were so pronounced that early geologists used them to designate breaks between eras. A close look at the boundary at the end of the Mesozoic (often called the K/T boundary for Cretaceous/Tertiary) illustrates some of the puzzles scientists have been encountering with these extinctions.

In 1980, a pair of scientists proposed that a large meteorite impact, throwing up dust and causing the world to become shadowed and cold, killed off the thousands of species eliminated by the K/T extinction. Deposits of iridium, an element recognized in many meteorites, are commonly found right at this boundary. A major meteorite crater discovered in subsurface records along the edge of the Yucatan Peninsula in Mexico was soon touted as the impact site. Still later, other scientists noticed that the fossil record revealed a major extinction about every 26 million years, and claimed a periodic extraterrestrial cause for the extinctions.

There are other twists to the scenario, however. Fossil evidence suggests that even before the end of Cretaceous time there was a gradual decline in the number of species; does this decline suggest that gradual climate change had more to do with the extinction than did the meteorite impact? Tropical and subtropical species seem to have been hit the hardest; does this support a climate-change cause, or could species living at higher latitudes be inured to cold and dark and therefore better able to survive meteorite impacts? Could there have been many meteorite impacts, a long-drawn-out shower of large meteorites, rather than just one? In India, massive flood basalts erupted at about the same time; did they release gases or dust that affected atmospheric quality around the world? Was the K/T extinction due to one cause, or did several catastrophes combine to alter the world in a way that made life inhospitable to many species?

somewhere else, there do seem to be sudden breaks in the orderly progression of plant and animal life. Some groups seem to go on forever, but others gradually or abruptly became extinct (see sidebar).

Breaks in the rock record, including those coinciding with time divisions of the geologic calendar, are called **unconformities.** They commonly take the form of erosion surfaces that bevel or channel older rock layers, with younger rock layers then deposited on top of them. Unconformities and the breaks in the fossil record that they entail are due to times of widespread uplift and erosion or to environmental changes. Unconformities occur on small scales too, representing short or localized intervals when sedimentation ceased and erosion held sway. Gaps in the rock record are commonplace; there is no place on Earth where we have a complete, uninterrupted record of all of geologic history.

VII. MAPS AND FIGURES

Photographs have become an integral part of geologic literature. They do not completely edge out maps, diagrams, and sketches, however, which sometimes do a better job of clarifying basic geologic relationships. Let's look at the types of geologic illustrations in common use.

Geologic maps show rock units—specific, named, easily recognizable units called **formations** or **groups** of formations—that occur at the Earth's surface or just under the loose soil and rock debris on the surface. Geologic maps are the prime product of most geologic field research, and result from plotting rocks visible in

Most of the Colorado Plateau is drained by its namesake the Colorado River and its tributaries, whose winding canyons produce much of the scenic beauty of this area.

outcrops or revealed by shallow digging, with help from aerial and satellite photography and other types of remote sensing. Geologic maps that correspond to topographic quadrangle maps can be obtained over the Internet or from U.S. Geological Survey map offices in Reston, Virginia; Denver, Colorado; and Menlo Park, California. Park and monument geologic maps can be purchased at some visitor centers.

Cross sections diagrammatically slice open the rocks below the surface to give a picture of what geologists think is there. **Block diagrams** combine cross sections with perspective views of surface features, showing what a block of the Earth's crust would look like if it could be lifted out of its natural surroundings. Cross sections and block diagrams are usually easier to understand than maps, and are good ways to show geologic features at a glance. In both, the vertical dimension may be exaggerated in order to show the succession of rock layers more clearly. By convention, unconformities are shown by wiggly lines.

On the facing page is another type of geologic illustration derived from cross sections: a **fence diagram.** Made from many cross sections put together in their correct geographic position, fence diagrams assemble a lot of data in one picture. This one shows the position, age, and general composition of rocks in national parks and monuments discussed in this book, and is a useful guide to the geologic history of the entire Plateau.

Another type of illustration particularly suited to the Colorado Plateau is the **stratigraphic diagram**—a picture summarizing the rock layers exposed in many outcrops, as they would appear if piled on top of one another in the order in which they were originally deposited. Stratigraphic diagrams can also show how rock layers weather and erode—some as slopes, some as ledges and cliffs—a feature that makes them extra useful on the Colorado Plateau, where a lot of scenery results from these very characteristics. Under natural conditions, individual rock layers may not everywhere erode in exactly the same way, so one must make some allowances in trying to match up scenery and stratigraphic diagrams.

VIII. THE PLATEAU STORY

With all these basic concepts in mind, we are ready for a quick overview of the geologic history of the Colorado Plateau. Let's take it era by era.

Precambrian Time. The history of the Plateau begins in Precambrian time, more than 1.7 billion years ago, when ancient gneiss and schist now visible in the depths of the Grand Canyon and in the Black Canyon of the Gunnison were formed. These rocks came into being when yet older sediments and volcanic rocks of island arcs and one or more small continents collided with and were accreted by a larger ancestral continent, vastly increasing its size. Subjected to extreme temperatures and pressures, the accreted rocks were tightly

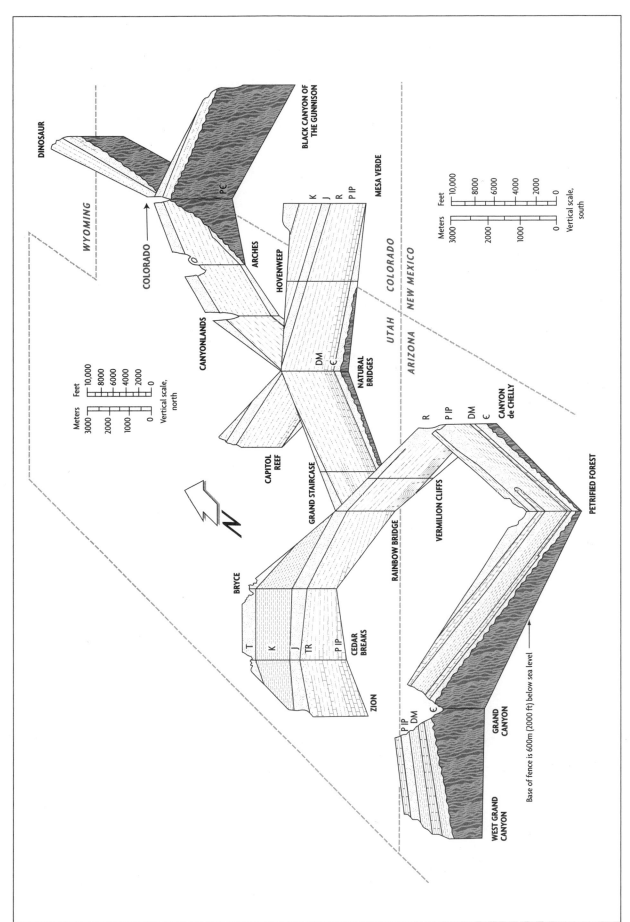

A fence diagram illustrates the Paleozoic, Mesozoic, and Cenozoic sedimentary rocks that appear in parks and monuments of the Plateau country. For abbreviations, see geologic time chart.

In the Inner Gorge, the Kaibab Trail descends through 1000 feet (300 meters) of hard, distorted Precambrian metamorphic rocks to cross the Colorado River footbridge. Beveled by erosion at the end of Precambrian time, the metamorphic rock is overlain by horizontal Cambrian strata. Ray Strauss photo.

folded, steeply tilted, partially recrystallized, and intruded by granite magma derived from the continent itself.

Most intrusion of granites ended by about 1.4 billion years ago, and the whole landscape was gradually eroded to a nearly horizontal surface. This surface was later—yet still in Precambrian time—overlain by younger sedimentary and volcanic rocks. Eventually these layers were broken and tilted by block faulting; then they, too, were beveled during the last 300 million years of Precambrian time. Few remnants of them remain in the Plateau region today; where they occur, they are still clearly recognizable as conglomerate, sandstone, shale, and dark lava flows.

Paleozoic Era. The early part of the Paleozoic Era was a time of comparative calmness and sameness. The beveled land rose and fell several times, just enough to allow a shallow western sea to advance and retreat repeatedly across the nearly horizontal land. In this area, early Paleozoic marine deposits on the whole were thin—several hundred feet during Cambrian time, little or none (none are preserved, anyway) during Ordovician and Silurian time, next to none during Devonian time. About 500 feet of limestone were deposited in the Mississippian sea.

The shallow Paleozoic seas were lively places. Sea-dwelling animals that had their beginnings well back in Precambrian time drifted and crawled and swam and burrowed in the soft sediment of the seafloor: brachiopods, trilobites, corals, clams and snails, feathery bryozoans, and fishes, the first vertebrates. Beginning in Cambrian time, plant and animal shells added significantly to the sediments, building layers of limestone. Preserved as fossils, the shells and skeletons now give us glimpses of evolving life. By Devonian time,

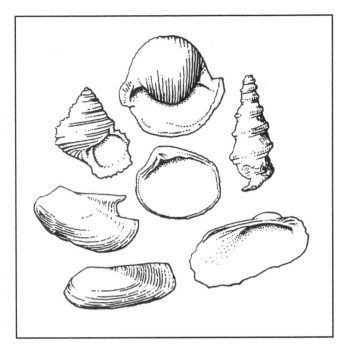

Fossils of many kinds are found in Plateau country rocks. These snails and clams are from the Permian Kaibab Limestone.

plants and some creatures of the sea had crept onto the land. In Permian time their descendants left footprints in damp dune sands, for the region still lay close to sea level, with shallow marine sediments alternating with delta, river floodplain, and dune deposits. Mountains that formed in Colorado, northern New Mexico, and central Arizona in Pennsylvanian and Permian time shed increasing amounts of sediments onto the flat Plateau region.

Mesozoic Era. With scarcely an interruption, the land rose near the end of Paleozoic time, and the western sea drained away. Sediments from the mountains continued to build floodplains and deltas. Now and then, nearby volcanoes belched volcanic ash.

Well to the west, subduction continued as arcs and microcontinents plowed into the edge of the continent, piling up mountains that may have waylaid moist winds from the sea. A great Sahara developed across the Plateau region, its high-piled dunes giving us many of the scenery-making rocks of the region. Just as today's Saharan dunes creep across the Nile floodplain, the dunes of this ancient Sahara advanced across older floodplains, covering and in places distorting, with their weight, the soft, slippery river muds.

Across both desert and delta wandered the newly dominant reptiles, among them the dinosaurs, for nearly 200 million years the monarchs of the land. In forests and swamps, trees lived and died, some perhaps perishing in showers of volcanic ash and great mudflows resulting from volcanic eruptions. And as a shallow sea again swept this area, large marine reptiles and elegantly spiraled ammonites,

Horizontal cliffs, ledges, and slopes stripe Grand Canyon's walls. Paleozoic rocks in the canyon range from Cambrian to Permian. Halka Chronic photo.

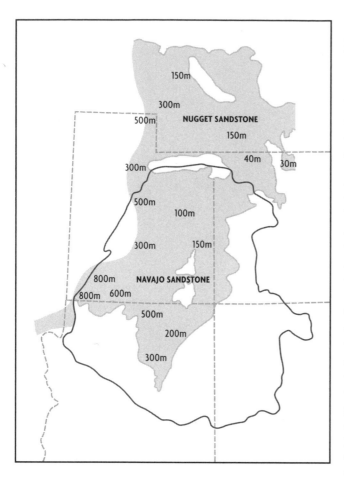

The Navajo Sandstone and its northern equivalent, the Nugget Sandstone, were deposited on a Triassic–Jurassic desert comparable to today's Sahara. Both formations thicken westward.

shelled relatives of squids and octopuses, swam above the fine gray mud of the seafloor.

Toward the end of Mesozoic time, North America, powered by the slow roll of convection in the Earth's mantle, broke away from Europe. As the newborn Atlantic Basin widened and westward movement of the continent increased, subduction to the west increased as well, its effects reaching farther and farther inland, reactivating old faults, lifting the Plateau and adjacent areas, and finally raising the Rocky Mountains. The Rockies would provide headwaters for the great river that, with its tributaries, would carve and chisel the scenic Plateau country we know today.

The end of the Mesozoic Era, perhaps punctuated by meteor showers and asteroid impacts that brought about a long night of airborne dust, saw the extinction of more than half of the plant and animal species then in existence, including the last remaining species of dinosaurs.

Cenozoic Era. With the disappearance of the dinosaurs, animals that survived the great extinction spread and diversified. In Tertiary time, ancestors of today's horses, elephants, pigs, and camels roamed this

region, feeding on the lush growth of a savanna environment. They in turn became prey to evolving members of the dog and cat families.

Erosion of newborn mountains surrounding the Plateau region swept rock debris into deep basins between mountain ranges. Soon the entire area was covered with a sea of sediment. Volcanoes were widespread, contributing significant amounts of volcanic ash to the deposits.

As the continent continued its western movement, overriding the Farallon Plate, it reached and overrode the Mid-Pacific Rise as well. It's not certain that the bulge and heat output of the Mid-Pacific Rise influenced the Plateau region directly, but stretching of the crust as it crossed that bulge may have been instrumental in creating the Basin and Range region to the west.

In Miocene-Pliocene time, several things happened that did influence the Plateau region. A multistate area centered on the Rocky Mountain region lifted, with uplift totaling 5000 feet or more. Faults along the western edge of the Plateau became active at about this time, dropping the Basin and Range region relative to the Plateau. The Pacific Plate's northwestward movement strengthened movement along the edge of the continent that continues today along the San Andreas Fault. The Plateau region rotated clockwise and tilted very slightly to the west. And 6 to 10 million years ago the Gulf of California opened up, a long narrow seaway that, until it was filled in, extended much farther north than it does today.

With an elevation difference of thousands of feet between the Plateau and the Basin and Range region, erosion increased, and a new drainage pattern was established, with a large river from the Rockies—the

Badlands of Triassic and Jurassic mudstones characterize many parts of the Plateau. Halka Chronic photo.

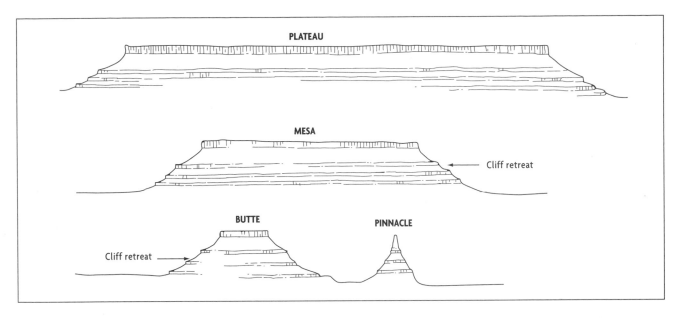

PLATEAU

MESA

← Cliff retreat

BUTTE

PINNACLE

Cliff retreat →

Most Plateau country erosion involves the retreat of slopes and cliffs, as plateaus reduce to mesas and mesas reduce to buttes and pinnacles.

Colorado River—ultimately finding a route to the Gulf of California.

On the Plateau, layers of hard sandstone or limestone alternate with softer siltstone or mudstone. As a result, most of the erosion occurs when cliffs of harder rock are undermined by the washing away (or blowing away) of softer layers just below, creating the alternating cliffs and slopes so characteristic of this region. Because the rocks dip gently northward over most of the Plateau, most of the cliffs gradually retreat in that direction.

As the Plateau developed, molten magma squeezed up along fissures and faults to create a number of domed laccoliths. Other magma burst through to the surface, building into several large composite or stratovolcanoes, among them San Francisco Mountain and Mount Trumbull. Later, as the Colorado River and its tributaries fashioned today's drainage patterns, cinder cones and dark basalt lava flows added to the contrasts of Plateau country color. Some lava flows dammed the rivers, even the mighty Colorado, creating temporary lakes or bringing about changes in their courses.

One of the last eruptions on the Plateau, that of Sunset Crater, occurred about 900 years ago. In all likelihood, there will be other eruptions; the region is by no means "dead" volcanically.

The Ice Ages of Pleistocene time also helped to shape scenery here. A few mountain glaciers reshaped the highest peaks, carving scoop-shaped cirques and leaving rough piles of glacial debris at their lower ends. Rainy cycles in phase with glacial advances elsewhere certainly added to the strength of the rivers carving the Plateau's many canyons.

More recent changes in this region have been brought about by humans, who have created dams and lakes where once there were canyons. Humans, too, have cut forests and grazed cattle and sheep in this country, reducing vegetative cover and adding to the erodability of the land. In places the earth has been gouged for minerals, and a thickening pall of pollution reduces the clarity of the once pristine air.

Small maps distributed at entrance stations by the National Park Service will help you find your way around in the parks. More detailed topographic maps and, for some parks, geologic maps, are available at visitor centers and on the Internet.

IX. OTHER READING

Most of the books and articles listed below cater to readers with little or no background in geology. The list does not include the many excellent and up-to-date textbooks now available at both high school and college levels, or professional journals and publications. Specific sources are listed with individual parks and monuments. Many other sources are now referenced on the Internet.

Baars, Donald L., 1995. *Navajo Country: A Geology and Natural History of the Four Corners Region.* University of New Mexico Press.

——, 2000. *The Colorado Plateau: A Geologic History.* University of New Mexico Press.

——, 2002. *A Traveler's Guide to the Geology of the Colorado Plateau.* University of Utah Press.

Chronic, Halka, 1983. *Roadside Geology of Arizona.* Mountain Press Publishing Co.

——, 1987. *Roadside Geology of New Mexico.* Mountain Press Publishing Co.

——, 1990. *Roadside Geology of Utah.* Mountain Press Publishing Co.

———, and Felicie Williams, 2002. *Roadside Geology of Colorado*, 2nd ed. Mountain Press Publishing Co.

Fillmore, Robert, 2000. *The Geology of the Parks, Monuments and Wildlands of Southern Utah.* University of Utah Press.

Hopkins, Ralph Lee, 2003. *Hiking the Southwest's Geology: Four Corners Region.* Mountaineers Books.

Luchitta, Ivo, 2001. *Hiking Arizona's Geology.* Mountaineers Books.

Nations, Dale, 1997. *Geology of Arizona.* Kendall/Hunt Publishing Co.

Powell, John W., [1885] 1995. *The Exploration of the Colorado River and Its Canyons.* Dover Publications.

Wiewandt, Thomas, and Maureen Wilks, 2001. *The Southwest Inside and Out: An Illustrated Guide to the Land and Its History.* Wild Horizons Publishing.

PART 2

THE NATIONAL PARKS
AND MONUMENTS

ARCHES NATIONAL PARK

ESTABLISHED: 1929 as a national monument, 1971 as a national park
SIZE: 125 square miles (323 square kilometers)
ELEVATION: 3900 to 5700 feet (1195 to 1723 meters)
ADDRESS: PO Box 907, Moab, Utah 84532

STAR FEATURES

- Stone arches—more than 200 of them discovered and documented—as well as windows, arch-shaped alcoves, and mazes of mini-canyons between tall fins of pink sandstone.
- Red and pink rocks deposited in river floodplains, on ancient dunes, and in near-shore lagoons.
- Unusual anticlines caused by the upward movement of underground salt, and valleys resulting from their collapse.
- Dinosaur tracks, part of an extensive, much-tracked area.
- Visitor center with geologic exhibits, a self-guided road tour, many trails (some with guide leaflets), guided tours in the summer describing geology, biology, and history.

See color pages for additional photographs.

SETTING THE STAGE

In eastern Utah, near the northern limits of the Colorado Plateau, flat-lying layers of sedimentary rock are corrugated by several northwest-trending anticlines. In Arches National Park, coral-colored sandstone of the Entrada Formation has responded to the collapse of two of these anticlines by splitting along parallel, northwest-trending, vertical joints. The joints widen as rain and snowmelt dissolve the calcium carbonate that binds the sand grains together. Soil develops along the joints, and plants find footholds, furthering the dissolving process with products of their metabolism and decay. Seasonal freezing and thawing, rainwash, and wind widen the joints until the rocks stand as tall, slender fins.

As erosion continues, some parts of the rock fins weather faster than others, and alcoves and caves form as thin, curving sheets of rock fall away. Where such localized weathering attacks both sides of a fin—perhaps along a zone of weakness in the rock, perhaps where underlying mudstone channels moisture—caves may develop on both sides of a fin. Deepening, they

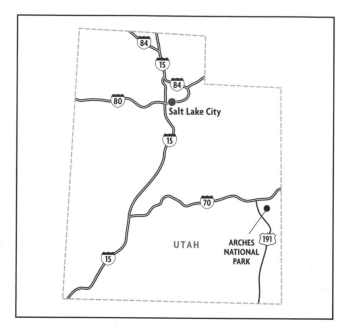

The ultimate in arch formation, Landscape Arch is 287 feet (88 meters) long and 106 feet (32 meters) high. Yet it is only 10 feet (3 meters) thick at its narrowest point. Tad Nichols photo.

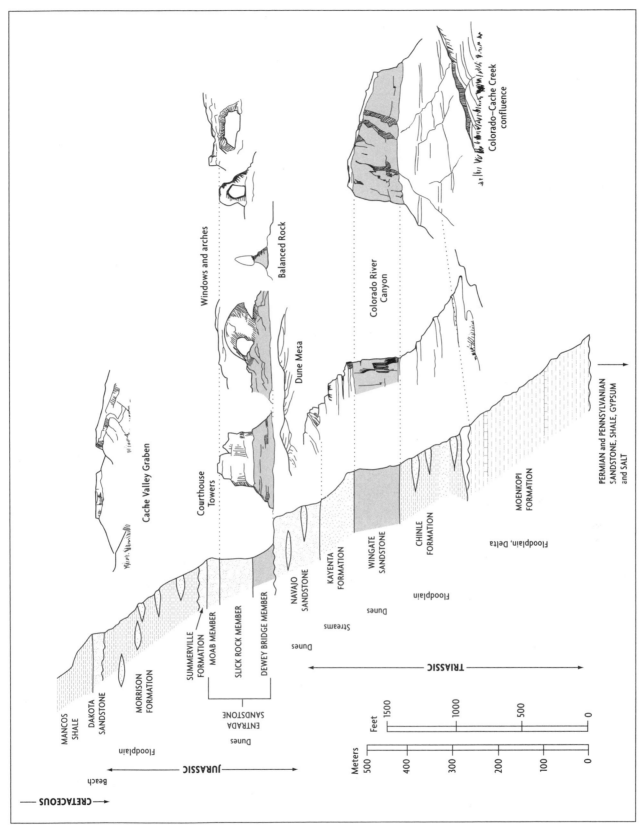

Stratigraphic diagram of Arches National Park

ultimately join, creating a window or an arch. Other arches form where solution works downward and then sideways from a pothole on the upper surface of a butte.

With time and further erosion, arches become increasingly slender and fragile. Eventually, unable to support their own weight, they collapse into piles of rubble. But arch formation is a continuous process, going on today: Fins are narrowing, new arches are forming, old ones are collapsing, and fallen rubble is being washed and blown away.

The secret to arch formation here is the presence of massive sandstone layers with abundant parallel vertical joints caused by bending across the Salt Valley and Cache Valley anticlines. The anticlines themselves are unusual: Cores of grayish and yellowish gypsum are exposed in valleys along their crests. These valleys commonly also contain tilted, irregular masses of rock younger than those exposed on their walls, suggesting that they were produced by collapse.

But what caused their collapse? The answer will surprise you: salt. Well below the surface here are thick layers of salt, gypsum, and potash originally deposited as sea water evaporated in Pennsylvanian time. When deeply buried, these evaporite minerals become plastic, able to flow slowly like Silly Putty or glacial ice. Just as Silly Putty will flow out from under your fingers if you press on it steadily, the salt and gypsum tended to flow toward areas where overlying strata were slightly thinner and pressures therefore not quite as intense—in this region, along underground ridges that mark ancient northwest-trending faults.

Less dense and more buoyant than overlying sedimentary rocks, salt and gypsum also pushed upward, exerting enough upward pressure to arch overlying rock layers into anticlines that inherited the northwest-southeast orientation of the ancient faults. Eventually most of the salt was dissolved away; only the gypsum now remains in the cores of the anticlines.

GEOLOGIC HISTORY

Precambrian Time. Although no Precambrian rocks are exposed within Arches National Park, their presence far below the surface has, as we have seen, influenced the geology and therefore the scenery here.

The Precambrian rocks are probably like those exposed in the Uinta Mountains farther north, in the Black Canyon of the Gunnison, and in Colorado National Monument: gneiss, schist, and granite at least 1.4 billion years old. Late in Precambrian time, these rocks were faulted into long northwest-trending ridges.

Paleozoic Era. During most of Paleozoic time, this region, along with practically all of the western United States, was low, almost flat, and often submerged by the sea. About 2000 feet (600 meters) of marine sedimentary rocks were deposited here. In Pennsylvanian time a platform in south-central Utah and highlands along the Colorado-Utah border, the latter an Ancestral Rocky Mountain range known as Uncompahgria, restricted the easy flow of sea water. In a subsiding area between the Utah platform and Uncompahgria, as the minerals in the water were concentrated by evaporation, great thicknesses of salt and gypsum precipitated and sank to the bottom. Thinner bands of black shale, usually taken as indicators of stagnant conditions, accumulated as well.

During Permian time these strata were covered with marine sediments and with rock debris washing from Uncompahgria. And the salt and gypsum that accumulated in Pennsylvanian time began to flow very, very slowly toward the old Precambrian ridges, and upward above them, pushing up low salt-cored anticlines.

Mesozoic Era. For most of this era this region was above the sea. Triassic rocks record the retreat of the sea in their change upward from marine limestone to brilliant red shale and sandstone. The Moenkopi Formation consists of sandstones, mudstones, and shales made up of debris washed from Uncompahgria and

FOOTPRINTS IN THE SAND

In the Klondike Bluffs area of Arches National Park and in other exposures outside the park, the Entrada Formation contains what may amount to literally millions of dinosaur tracks. Individual footprints range from about 10 to 20 inches (25 to 50 centimeters) in length. As far as is known, the great creatures that walked this way left no bones behind for scientists to study, but the footprints indicate that they may have been ancestors of Allosaurus.

Paleontologists surmise that the footprints, in a dense swath that continues for many miles, were made during a pause in the deposition of the sandstone. For some time a wet, sandy surface remained fairly soft, and the tracks themselves were impressed by individuals or small groups repeatedly crossing the same territory, rather than by a single immense herd of dinosaurs. Lying on an almost horizontal surface, most of the exposed footprints are now rapidly eroding.

The tracks occur near the top of the Entrada Formation, in the Moab Member, not far below the softer Summerville Formation. The Summerville is stratigraphically just below the Morrison Formation, known at other localities to contain numerous dinosaur fossils. All three formations are Jurassic.

The Egyptian queen and other figures result when jointed sandstone layers are undermined by erosion of soft shale and mudstone like that in the foreground. The queen's head is thought to have been offset by an earthquake. National Park Service photo.

other highlands and deposited on river floodplains, on mudflats, and in shallow bays. The brightly and variably colored Chinle Formation, forming a colorful slope below the cliffs of the Wingate Sandstone, consists of interbedded shale and siltstone, with volcanic ash contributed by volcanoes erupting well to the west and southwest.

Late in Triassic time, and continuing into Jurassic time, desert conditions prevailed and sand dunes spread across much of the western interior. Both the Wingate and Navajo Sandstones bear the sweeping diagonal cross-bedding of the dunes, as does the Entrada Sandstone that forms the fins and arches. Dinosaur tracks appear in some of the layers of sandstone (see sidebar). Above the dune sandstone, mostly eroded from this national park but visible to the north of it as well as in Cache Valley graben, are widespread sheets of rainbow-hued mudstone, siltstone, conglomerate, and limestone of broad river floodplains and freshwater lakes, making up the Morrison Formation.

Double Arch formed, like many other arches, as erosion of the crumply Dewey Bridge Member undermined the stronger sandstone above. Halka Chronic photo.

Upward movement of Pennsylvanian salt continued sporadically through Triassic and early Jurassic time. Though many of the Triassic and Jurassic rocks thin and disappear near the salt anticlines, the uppermost Jurassic sediments at one time extended right across the anticlines, suggesting there was a pause in the upward push of the salt.

As the Cretaceous Period began, seas again advanced, coming this time from the east. Few Cretaceous rocks are preserved in Arches National Park; in neighboring areas they show a cycle of shore sand, marine shale, and shore sand again. The upper sandstones are interlayered with river and delta deposits and coal beds that tell us of lagoons and swamps along the Cretaceous shore. The Cretaceous rock sequence can be seen to the northeast from high vantage points near Devils Garden: Book Cliffs, the lower band of cliffs curving across the Colorado-Utah border, display soft Cretaceous marine shale capped with a sandstone-coal-sandstone sequence. At the end of Cretaceous time, the sea withdrew.

Cenozoic Era. In late Cretaceous and early Tertiary time a wave of mountain building swept eastward from California and Nevada, buckling parts of eastern Utah and western Colorado and thrusting up the Rocky Mountains. The Arches area was caught between the newly formed San Rafael Swell of central Utah and a new Uncompahgre Uplift in much the same position as the earlier Uncompahgria. The Arches area seems to have bowed downward, though not as sharply as regions farther west and north, where particularly thick lake sediments accumulated, some of which appear as Roan Cliffs, the upper range of cliffs visible to the north from high points within Arches National Park.

All during Paleozoic, Mesozoic, and Cenozoic time there were sporadic movements on the old Precambrian faults that underlie this region and control the position of the salt anticlines. Movement of the salt and bowing up of the salt anticlines are believed to have occurred in two phases. The first phase began in Permian time and continued until Jurassic time—a very slow, perhaps pulselike process. A second phase in Tertiary time intensified the shape and size of the anticlines and led to their ultimate collapse.

In Miocene-Pliocene time, the whole region—Utah and Colorado and parts of adjacent states—rose 5000 feet (1500 meters) or so as a broad, almost indiscernible dome, bringing this region to its present elevation. Though many surface features remained nearly the same, uplift increased rates of erosion. Rivers and streams, steepened and strengthened by uplift, bit into rock layers along their paths, carving the many splendid canyons of the Plateau country.

With this region well above sea level, groundwater dissolved and removed the salt from the salt anticlines, flushing it away into the Colorado River. As salt

was removed from the anticlines, their crests collapsed, dropping in some cases many hundreds of feet. Each collapse carried down with it some of the younger rock that surfaced this area at the time. Each is now marked by long fault-edged valleys floored with gypsum, not nearly as soluble as salt, and fragments of younger rocks, bordered with eroded bluffs, towers, and arches of sandstone.

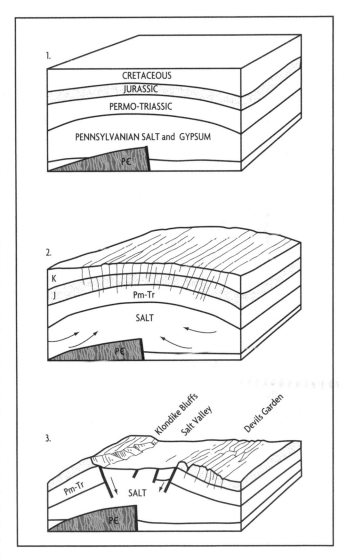

Evolution of the Salt Wash Anticline:
1. *Thick layers of salt overlie an ancient fault. They are blanketed by hundreds of meters of Permian and Mesozoic rocks. Because of continued movement along the fault, sediments thin above it.*
2. *The weight of overlying rocks forces salt to flow upward toward the ancient fault, the area of least overburden, doming up and cracking overlying sedimentary rocks. Continued fault movement emphasizes the fold.*
3. *Removal of part of the salt by groundwater causes collapse of the anticline crest. Cretaceous and Jurassic rocks drop to form the valley floor, where gypsum remains as gray hills.*

Gradually, vast upland areas were swept clean of soft, easily eroded Tertiary deposits. Frost wedging, resulting from sharp temperature changes, loosened sand and rock fragments, which then were swept away by wind and rain. Though they brought no ice to this immediate area, the Ice Ages of Pleistocene time did bring cooler temperatures and probably greater precipitation, undoubtedly augmenting erosional processes. Streams and rivers became roaring floods, scouring ever-deeper channels and sweeping ever more sand and silt from the highlands, creating the scenery we see today.

BEHIND THE SCENES

Cache Valley Graben. One of the park's two collapsed salt anticlines, Cache Valley can be seen from viewpoints along the main park road or visited by the road to Wolfe Ranch. Bordering rock layers, which dip away from the valley, are the sides of the anticline. The crest of the anticline collapsed as its salt core was dissolved away, bringing down with it some of the rocks that covered the surface at the time of its collapse: Cretaceous Mancos Shale and Dakota Sandstone.

Bright green rock near the Wolfe cabin contains glauconite, a mineral thought to have been deposited in agitated water. Green marble-sized balls formed as concentric layers of glauconite were deposited on sand grains rolled to and fro by waves.

Surface water here is strongly flavored with salt and gypsum. Springs in recesses below the highest cliffs, which provided the Wolfe family with palatable drinking water, show that groundwater is moving slowly through the porous sandstone, especially just above an impermeable shale layer in the Entrada Sandstone.

Indian petroglyphs on the east side of Salt Wash, a little north of the Wolfe Ranch, were pecked through desert varnish that coats a sandstone slab, revealing the lighter rock below. Some were carved in historic time after horses were brought to the Americas by Spanish conquistadores. Even though the petroglyphs have been here for many hundreds of years, they show no trace of new desert varnish.

Colorado River Canyon. It's hard to get directly to the Colorado River from Arches National Park, but by going upriver from Moab you can enjoy a scenic drive and look across the river into the south edge of the park.

Rocks exposed along the river are older than those elsewhere in the park. The oldest are red sandstone and shale of the Moenkopi and Chinle Formations, of Triassic age. Above them rise high angular cliffs of Wingate Sandstone, Triassic dune sandstone, capped by thin, resistant ledges of the Kayenta Formation. Lighter-colored, rounded cliffs along the skyline are Navajo Sandstone, dune-formed on a Jurassic desert. This rock

Courthouse Towers are isolated fins of the Slick Rock Member of the Entrada Formation undermined by erosion of the Dewey Bridge Member. National Park Service photo.

Near Delicate Arch, fine laminations of wind-deposited sand festoon bare rock surfaces. Loose sand is swept away by wind. Ray Strauss photo.

forms the surface in the southern part of the park.

As you drive along the river, see if you can distinguish on the opposite bank the Courthouse syncline, the Salt Wash anticline (where Salt Wash Creek enters the Colorado), and the highly faulted Cache Valley anticline. All these structures extend south across the river but peter out near the La Sal Mountains.

Courthouse Towers. Courthouse Towers and the tall cliffs nearby are remnants of rock layers exposed also in the Windows and Devils Garden parts of the park, the two lower members of the Entrada Formation. Here they occur in a gentle syncline, its lowest point marked by Courthouse Wash. The valley is floored with Navajo Sandstone. Ledges and slopes at the bases of the cliffs and towers are in the Dewey Bridge Member of the Entrada Formation, recognizable by its wavy, crumpled bedding.

Here, as elsewhere in the park, joints caused by formation and collapse of a salt anticline dictate the major patterns of erosion. Minor patterns are governed by differences in rock composition and hardness. As they break away along vertical joints, resistant sandstone layers form cliffs and ledges. Shale and mudstone form slopes, benches, and recesses that in some cases undermine the massive sandstone cliffs. Where undermined, the sandstone may break away in arched alcoves.

Delicate Arch. The trail to Delicate Arch offers some of the most delightful, most geologic vistas in the park, with chances to examine the lowest two members of the Entrada Formation. The contorted Dewey Bridge Member, near the trail beyond the footbridge, in places contains clumps and nodules of agate. This unit weathers fairly readily along mudstone beds, and commonly undermines the massive dune sandstone above.

Farther along, the trail crosses smooth, barren slopes of the Slick Rock Member, pale pink sandstone swept by sand-dune cross-bedding. Furrows have developed along joints in this rock, and little potholes line up along the furrows. After rains, these potholes hold small pools, homes for tiny short-lived animals and plants whose metabolic by-products add acid to the water, speeding solution of the calcium carbonate that binds the sand grains of the rock together.

Delicate Arch and its surroundings display well the subtle differences of color and resistance to weathering that characterize the Slick Rock Member. Many small, deep cavities, carved by wind, line up along particularly easily eroded beds. Once these hollows are deep enough to hold a few sand grains, the wind is assured of a supply of tools, and whirling eddies bombard the hollows with sand, rounding and deepening them.

Devils Garden and Fiery Furnace. Red sandstone fins that result from erosion along parallel joints are thinner, higher, and more numerous here than in other parts of the park. The Devils Garden Trail follows mini-canyons between the fins and finally climbs to the crest of one of them. Many arches are hidden in this rocky labyrinth, a number of them visible from this trail. All are different, and here you can pinpoint the factors that led to their development: erosion of the weak Dewey Bridge mudstone, closely spaced fracturing, weathering accelerated by the high moisture content of the rock, weathering along weak beds within the Slick Rock Member. Some arches are small and probably quite young. Others, like Landscape Arch (which may be the world's longest natural span), are about as large as they can get without collapsing.

Arch formation goes on today as it has for thousands of years: a sand grain or a pebble or a rock slab at a time. Skyline Arch, visible from the road, more than doubled its size in 1940 when a large block of sandstone fell from it, but most changes are not that sudden and dramatic. On many surfaces the rock peels off in flakes because temperature changes bring about small fractures parallel to the rock faces, and water freezing and thawing in the cracks flakes off curving sheets of rock. Many smoothly contoured surfaces are shaped in this manner, and many rounded arches are enlarged in this way.

Entrance Road. Geology around the visitor center is complicated by the Moab Fault and the Moab Valley anticline, one of the northwest-trending anticlines of eastern Utah. The valley is lopsided: Rocks on its northeast side, behind the visitor center, are of very late Triassic and Jurassic age, younger than the Permian and Triassic strata across the valley. Relative displacement between the two sides of the valley is about 2600 feet (800 meters).

Leaving the visitor center, the road climbs through Jurassic strata—a good place to get acquainted with them, as exposures are top-notch and there are several turnouts for parking. The light buff or white Navajo Sandstone originated as sand dunes and interdune flats; it bears the long, sweeping laminations of the dunes and the silty horizontal beds of interdune deposits. Above it is gnarled red sandstone of the Dewey Bridge Member of the Entrada Formation. The Three Penguins are in the middle part of the Entrada Formation, the Slick Rock Member. This rock, too, is dune sand, and it is the layer in which are carved the arches of this national park. Several faults, with displacements up to about 50 feet (15 meters), are exposed on cliff faces near the road.

Beyond Courthouse Wash are many excellent exposures of the upper surface of the Navajo Sandstone, again showing the broad cross-bedding that characterizes dune sandstone.

Klondike Bluffs. These bluffs form the west limb of the Salt Wash anticline. They exhibit many sandstone fins and quite a few arches, among them Tower Arch with its stunning setting and lovely view. Along the trail to Tower Arch, many characteristics of the rocks that form the fins and arches are displayed: the knobby, contorted slopes and "goblins" of the Dewey Bridge Member; the massive, cross-bedded, cliff-forming sandstone of the Slick Rock Member; and the thin-bedded white sandstone of the Moab Member, all parts of the Entrada Formation. Watch for examples of desert varnish, prominent diagonal layering of wind-deposited (and wind-eroded) sandstone, honeycomb weathering, and exfoliation, a form of weathering in which thin curving slabs peel off rock surfaces. Parallel joints control fin development here, as on the northeast side of Salt Valley.

Park Avenue Trail. The contorted shale and mudstone of the Dewey Bridge Member and the hard, resistant, cliff-forming sandstone of the Slick Rock Member of the Entrada Formation are well exposed here. Notice the differences between ledge-forming sandstone and slope-forming shale and mudstone. Contorted bedding in the Dewey Bridge Member is thought to have originated before the mud of which it was formed consolidated into rock—when it was soft enough to slump and slide. Some slumping may have been caused by the weight of overlying sediments, but the contortions do not extend up into the massive sandstone.

Desert varnish stains many cliffs above the trail.

Salt Valley. At the heart of Arches National Park, Salt Valley is a collapse valley bordered by bluffs and cliffs of Jurassic sandstone. The road into it follows a creekbed through steep gray and yellow gypsum hills that are clues to the origin of the valley. When the crest of the Salt Wash anticline collapsed, it carried down with it both Pennsylvanian gypsum and large fractured blocks of overlying strata.

Windows Section. Differential weathering of the lowest two members of the Entrada Formation, and the roles they play in arch formation, are particularly well demonstrated in this part of the park. The crumpled mudstone and sandstone of the Dewey Bridge Member almost invariably weather and erode more rapidly than the massive Slick Rock Member above. Balanced Rock is a fine example of this: The balanced boulder of Slick Rock Sandstone perches on a pedestal of weaker Dewey Bridge mudstone. Both rise above a rolling surface of Navajo Sandstone, whose hard and soft layers weather in such a way as to emphasize its dune-derived cross-bedding.

Most of the arches here, including Double Arch, the North and South Windows, and Turret Arch, have also formed right at the Slick Rock–Dewey Bridge contact, where erosion of the mudstone undercuts overlying sandstone.

Pothole Arch, high up on the northeast side of

Elephant Butte, developed in a different manner: A pothole on the upper surface of the butte deepened as rainwater pools, made acid by the metabolism of tiny aquatic plants and animals, dissolved the sandstone's calcium carbonate cement. At about the same time, a more or less horizontal recess formed in the wall of the butte along a weak layer in the sandstone. As erosion continued, cave and pothole met.

OTHER READING

Baars, Donald L., 1994. *Canyonlands Country: Geology of Canyonlands and Arches National Parks.* University of Utah Press.

Johnson, David W., 1985. *Arches: The Story Behind the Scenery.* KC Publications.

Schneider, Bill, 1997. *Exploring Canyonlands and Arches National Parks.* Falcon Publishing Co.

BLACK CANYON OF THE GUNNISON NATIONAL PARK

ESTABLISHED: 1933 as a national monument, 1999 as a national park
SIZE: 32 square miles (84 square kilometers)
ELEVATION: 5440 to 8561 feet (1658 to 2609 meters)
ADDRESS: 102 Elk Creek, Gunnison, Colorado 81230

STAR FEATURES

✪ The deep, narrow canyon of the Gunnison River, carved in ancient rocks that form the foundations of the continent.

✪ An eventful canyon history involving mountain building, submergence and sedimentation, volcanism, and erosion.

✪ Visitor center, viewpoints, trails along the rim (some with trail leaflets), and, in summer, guided walks and evening programs.

SETTING THE STAGE

The Black Canyon's name can be credited to its shaded depths, where the sun rarely shines, and to the dark rocks of its precipitous walls. In many places the canyon is deeper than it is wide. Here are some vital statistics:

Total length: 53 miles (85 kilometers)
Length in national park: 14 miles (22.5 kilometers)
Greatest depth: 2722 feet (829 meters)
Average depth: 1968 feet (600 meters)
Width of river at Narrows: 40 feet (12 meters)
Width at rim above Narrows: 1300 feet (400 meters)

The walls of the canyon are composed of dark Precambrian gneiss and schist and lighter-colored granite. Veins of coarse-grained pegmatite slash obliquely across many parts of the canyon walls. Vertical joints form pathways for weathering and erosion and, as water seeps into them and freezes and thaws, contribute to the ruggedness of the canyon walls.

The Gunnison River, its headwaters high in Colorado's Rocky Mountains, carved this deep chasm with the torrents of its seasonal floods, when the untamed river battered rock with rock and swept debris downstream. Much of the carving dates to Pleistocene Ice Age time, when mountain glaciers released plentiful supplies of water and rock-fragment tools. In historic time the river has carried as much as 12,000 cubic feet (350 cubic meters) of water, sand, silt, pebbles, and boulders per second past any given point.

The river's steep gradient gave it the speed, power,

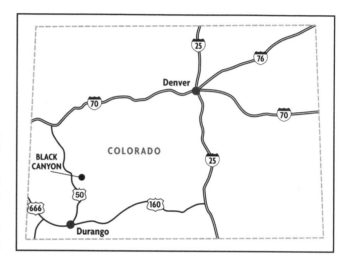

and turbulence to cut downward faster than other processes could flare out its rims. Widening of the canyon is the work of frost action, rockfall, and landslide, with the river carrying away the fallen debris.

These processes continue today, but now upstream dams hold back floodwaters, so the river is unable to carry away the debris of passing years. Few tributaries enter the Black Canyon as it cuts across the high central core of the Gunnison Uplift, but those that do meet the main canyon high up on its walls because they lack the cutting power of the larger river and cannot deepen their channels as effectively. In some parts of the canyon, lines of weakness such as joints, dikes, faults, and parallel mineral grains in schist provide avenues of drainage.

From viewpoints along the South Rim Drive, you can peer down into the somber chasm and see the hard crystalline rocks of its walls. These ancient rocks are extremely resistant, extremely tough, capable of standing in near-vertical walls well over 2000 feet high. The rock units themselves are steeply tilted, in places almost vertical. Within the national monument they line up in broad bands that trend northeast-southwest, at right angles to most of the canyon, like books on a bookshelf waiting to be read. And the story they tell is of the birth of the North American continent.

Near Gunnison Point and the visitor center, dark layered gneiss appears in the canyon's shadowy depths, some of it grainy and very like quartzite, some slaty or schistlike. Both types were baked and hardened, though never really remelted; they retain enough of

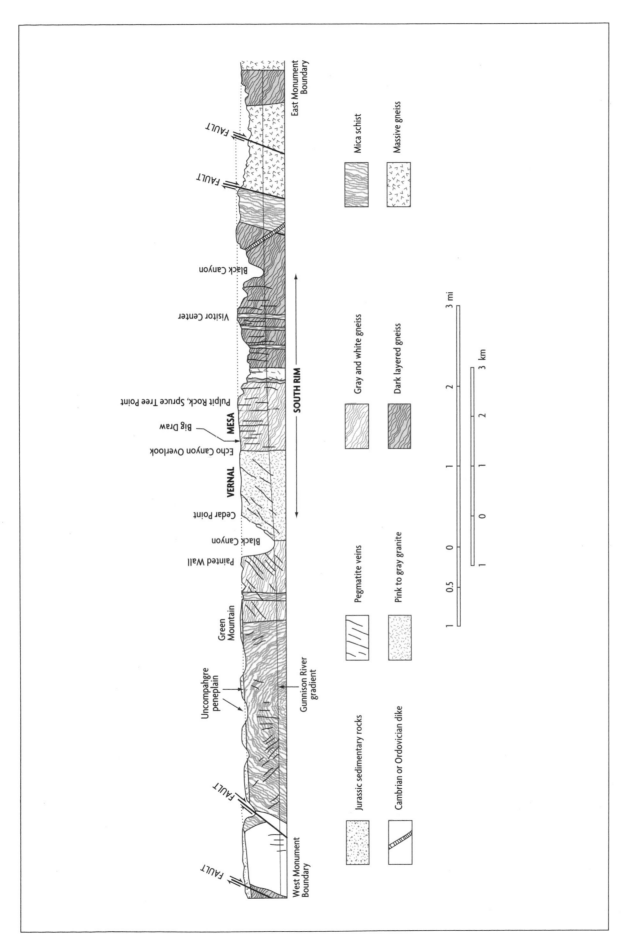

Ancient Precambrian rocks exposed in Black Canyon's walls are similar to those of Colorado National Monument and Grand Canyon's Inner Gorge. They range in age from 1.75 to 1.4 billion years old. Because of the river's winding course, this section crosses the river twice; the south rim shows in the center section.

their former characteristics to show that they were derived from layers of sandstone, shale, and siltstone. Some are still sandy in appearance, but with grains so tightly cemented that the rock breaks through the grains rather than around them—a characteristic of quartzite. Some contain scatterings of pebbles, evidence that they were once conglomerate.

Halfway between the monument headquarters and Pulpit Rock (Spruce Tree Point), where South Rim Drive turns north, the ancient gneiss is decoratively swirled with white and gray bands, the light swirls with quartz and feldspar predominant, the dark ones with abundant biotite. Large inclusions of granite are surrounded by darker ribbons of much finer-grained rock. Pure white quartz veins and light-colored pegmatite veins show up too, the latter with unusually large crystals of quartz, feldspar, and muscovite (white mica).

The origin of the gneiss is puzzling. Some of it looks as though it was squeezed as a semifluid porridge between layers of solid or slightly mushy older, darker rock. But in places the darker rock appears to have squeezed between masses of lighter rock, reversing this pattern.

Radiometric studies show that the gneiss is about 1.7 billion years old. Radiometric dating gives only the recrystallization age, however; the original rocks of which it formed must have been older.

West of Echo Canyon Overlook the rock type changes, and the South Rim Drive crosses a broad belt of coarse-grained granite. In this rock, flat-faced pink and white feldspar crystals and chunky, glassy quartz grains are large and easy to distinguish. The pink color of the feldspar crystals results from long-term bombardment by radiation from radioactive minerals in the granite. Smaller crystals of dark magnetite and hornblende give it a salt-and-pepper look. Little green crystals, hardly more than specks, are epidote.

The pinkish color carries through within the canyon walls, making this rock easy to recognize even from a distance. Its radiometric age is 1.45 billion to 1.35 billion years old, when it intruded during a phase of widespread mountain building. It is older than the many coarse-crystaled pegmatite veins that cut through it. The Gunnison River runs right along the west boundary of the granite intrusion, at the foot of the Painted Wall, along a zone of weakness that seems to have guided its course. The edge of the intrusion continues southwestward and passes between High Point, at the end of the road, and Warner Point.

At the Painted Wall (the tallest cliff in Colorado) lies more squiggly-patterned gneiss, laced artistically with pegmatite veins. Because of the curve of the cliff, the veins appear to arch, but they are really vertical sheets with a northeast-southwest trend almost parallel to the cliff face.

Along the Painted Wall, vertical joints control erosion of deep parallel fissures. The remarkably flat upper surface of all the Precambrian rocks here is a surface that geologists call the Uncompahgre Peneplain. Back from the canyon rim, the peneplain is covered with flat-lying sedimentary rocks much younger and much weaker than the Precambrian rocks of the canyon itself.

Downcanyon from High Point and Warner Point is a fourth rock type: silvery mica schist. It appears in the center of an anticline and walls most of the western part of the canyon.

GEOLOGIC HISTORY

Precambrian Time. The Black Canyon's story begins far back in Precambrian time, around 1.8 billion years ago, with sedimentary and volcanic rocks of an island arc that lay south of a small continent or tectonic

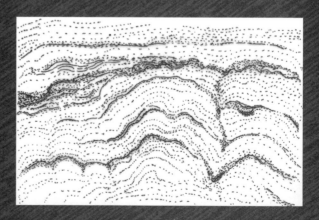

AN INHOSPITABLE WORLD

What was it like here in Precambrian time—say two billion years ago, when some of these rocks were forming? This area was an arc of islands about to crash into the Wyoming Province, part of the infant continent that would grow into North America. And the two plates—the one carrying the island arc and the one carrying the nascent continent—would collide with vigor, for in those days the skin of the Earth's crust was thinner and the mantle hotter, and tectonic plates moved at a faster pace.

You could not have "breathed easy," for the "air" around the globe was heavy with gases that would poison most life today. Oxygen, produced by simple, minute blue-green algae that were capable of doing without it, had reached about 15 percent of its current atmospheric levels. Tiny algae and bacteria ruled the seas, and algal mounds of stromatolites (see drawing) lived in favorable conditions in warm bays. On land no life existed, or could exist: the ozone layer was too thin to offer a shield against harmful radiation. Even in the daytime the sky was fairly dark, without the light-scattering oxygen we have today. And at night the moon, much closer to Earth, shone bigger and brighter, and pulled up tremendous tides on its journey around the barren planet.

Black Canyon's spectacular chasm, carved by the Gunnison River, is 1900 feet (570 meters) deep, yet the river's present channel is in places only 40 feet (12 meters) wide. Halka Chronic photo.

plate that geologists call the Wyoming Province (see sidebar). The island arc and the Wyoming Province collided around 1.7 billion years ago, creating tremendous pressures and temperatures, building mountains, thoroughly crumpling and crushing the rocks of the island arc, destroying their original layering, and at great depths reforming and realigning their minerals and turning them into dark, contorted gneiss and schist. As the altered rocks cooled and shrank, magma penetrated shrinkage cracks, leaving some of the dikes that now cut across the faces of the cliffs.

Around 1.4 billion years ago another round of tectonic action brought intrusion of granite, in places melting surrounding rock so that it was absorbed into the granite magma. As the rock cooled and contracted, fluid leftovers from crystallization flowed into cracks and fissures, where they cooled and crystallized slowly to form the giant crystals of pegmatite veins. Throughout the crystallization process, crustal movements kneaded the ancient dough and eventually folded it into tight anticlines and synclines with rock layers squeezed into vertical position. Later more dikes penetrated both the gneiss and granite.

Late in Precambrian time, continent-wide erosion planed away the land, making of it, as the era closed, an almost featureless peneplain quite close to sea level. The chances are good that the final shaping of this broad horizontal surface was the work of the sea.

Paleozoic and Mesozoic Eras. Soon after the Paleozoic Era opened, marine sediments began to accumulate on the beveled surface: thin sheets of limestone, shale, and sandstone, some entombing the shells of marine animals. Although the sea drew back at times, and some erosion occurred, sediments remained essentially horizontal until Pennsylvanian time, when the Ancestral Rocky Mountains—including, in this area, the range known as Uncompahgria—were thrust above the sea. As streams and rivers stripped away the easily eroded sedimentary rocks on their crests, they once more laid bare the old Precambrian erosion surface.

In Triassic time, this surface was covered with nearshore sandstone and lagoon mud and clay, sediments that now appear as the colorful strata eroded back from the northeast rim of the Black Canyon. Some layers contain limestone and gypsum, and suggest salty coastal ponds that dried up in an arid climate; some are of rainbow-hued mud and volcanic ash deposited on river floodplains. In Jurassic time, deserts of sand extended across this area.

The Cretaceous Period brought new incursions of the sea. Beach sand deposited along the shores of a retreating sea became in time the Dakota Sandstone, the resistant formation that now caps mesas near the national monument. During a second advance of the sea, the thick gray mud of the Mancos Shale accumulated in deeper water, mud now exposed west of the

Black Canyon in the fertile valleys of the Uncompahgre, Gunnison, and Colorado Rivers. Late in Cretaceous time the sea departed as the land began to rise with the onset of the Laramide Orogeny.

Cenozoic Era. The rise of the Rockies continued well into Cenozoic time as the North American Plate, having separated from Europe, drifted westward. The area around Black Canyon was broken once more along ancient faults, lifted, and tilted. New drainage patterns developed. At some time the ancestor of the Gunnison River was born, its course nearly the same as that of its present-day descendant.

Tertiary volcanism, however, gave the river a hard time. As clouds of volcanic ash burst from volcanoes north and south of the ancestral Gunnison, its course was pushed southward, then northward, then southward again as the stream sought the low points between

The Painted Wall exposes the many light-colored veins that ornament the Precambrian gneiss. The horizontal surface at the top of the photograph represents several long, superimposed periods of erosion. Halka Chronic photo.

two volcanic ranges, the San Juan Mountains to the south and the West Elk Mountains to the north.

As volcanism continued, the land around the volcanic centers sagged, perhaps compensating for the vast outpourings of lava and volcanic ash. This sinking, too, affected the river, especially a ring-shaped subsidence that partly encircled the West Elk Mountains. Around 2 million years ago, as volcanism abated and finally died away, the Gunnison found its present course along the southern arc of this depressed, ring-shaped syncline.

The river cut easily through volcanic tuff and breccia and Mesozoic sedimentary rocks, soon deepening its channel. Once the channel was deepened, the river could no longer change its course, which lay by chance above the old uplift of hard Precambrian rocks. When eventually the river cut down to the uplift, it could do little except continue to trench through the hard rock, carving out the canyon we see today. In this task it was helped by zones of weakness in the ancient rock: faults and joint planes and lines of contact between igneous and metamorphic rock.

Downward cutting was given a boost by uplift of this region in Tertiary time, and by changes in the course of the Colorado River, of which the Gunnison is a tributary. The rate of downward cutting must have varied as well with changes in climate, increasing, for instance, with swollen runoff and plentiful rock and sand tools during Pleistocene Ice Age time. To erode a canyon as deep as Black Canyon in, say, 5 million years, cutting would have to proceed at an average rate of about 4.8 inches (12 centimeters) every thousand years—only around a third of an inch (1 centimeter) in a human lifetime.

OTHER READING

Hansen, W. R., 1987. *The Black Canyon in Depth.* Southwest Parks and Monuments Association.

Houk, Rose, 1991. *Black Canyon of the Gunnison.* Southwest Parks and Monuments Association.

Lyons, Steve, 2002. *Black Canyon of the Gunnison Explorer's Guide.* Freewheel Publications.

Prather, T., 1999. *Geology of the Gunnison Country.* B&B Printers.

BRYCE CANYON NATIONAL PARK AND CEDAR BREAKS NATIONAL MONUMENT

BRYCE CANYON NATIONAL PARK
ESTABLISHED: 1923 as a national monument, 1924 as Utah National Park, 1928 as Bryce Canyon National Park
SIZE: 56 square miles (145 square kilometers)
ELEVATION: 6620 to 9105 feet (2018 to 2775 meters)
ADDRESS: PO Box 170001, Bryce Canyon, Utah 84717-0001

CEDAR BREAKS NATIONAL MONUMENT
ESTABLISHED: 1933
SIZE: 10 square miles (26 square kilometers)
ELEVATION: 10,350 feet (3155 meters)
ADDRESS: 2390 W. Highway 56, Suite #11, Cedar City, Utah 84720-4151

STAR FEATURES
- ✪ Two gigantic amphitheaters ornately decorated with colorful turrets, pinnacles, and freestanding walls carved as streams erode headward into two of the High Plateaus of Utah, the Paunsaugunt and Markagunt Plateaus.
- ✪ North-south faults that separate the plateau segments.
- ✪ Soft pink sedimentary rocks deposited in Paleocene and Eocene lakes and streams, including soil layers developed soon after they were deposited.
- ✪ Drives along the forested rims of the two plateaus, with viewpoints overlooking the eroded landscape.
- ✪ Many trails, some with guide leaflets, along the canyon rims. At Bryce, trails also lead down among the carved pink rocks.
- ✪ Visitor centers, museums, and, in summer, guided walks and evening programs.

See color pages for additional photographs.

SETTING THE STAGE
Along the east margin of the Paunsaugunt Plateau and the southwest tip of the Markagunt Plateau, mischievous nature has carved fanciful fairylands of elaborate cliffs and pinnacled ridges separated by deep, steep alleyways. The intricate, many-branched mazes are sculptured in soft pink rock of the Claron Formation, made up of Paleocene and Eocene lake and stream deposits that range from silty and sandy limestone to limy siltstone and sandstone, with liberal doses of clay and volcanic ash. The pink color of these rocks comes from minute amounts of iron oxides, mostly tiny grains

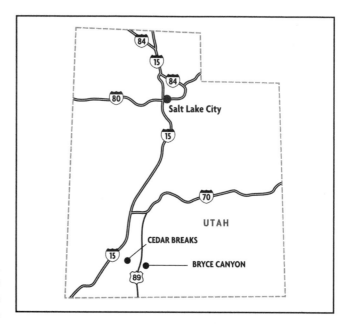

of hematite. At Cedar Breaks, the entire formation is considerably thicker, sandier, and siltier than at Bryce.

Below the lake deposits are older rocks—tan or gray Cretaceous sandstone—equally susceptible to erosion but tending to form ridged hills rather than steep gullies and ridges.

At both sites the Claron Formation is cut by faults. At Bryce, the Sevier Fault on the west, bordering the Sevier Valley, and the Paunsaugunt Fault on the east define the Paunsaugunt Plateau. At Cedar Breaks, the Hurricane Fault edges the Markagunt Plateau. Along all three of these faults, the east side is moved up relative to the west side.

The Sevier Fault is entirely outside the present park. It is exposed nearby, however, near the mouth of Red Canyon, where the pink rocks are pushed upward some 2000 feet (600 meters) and abut basalt dated at 500,000 years old, the fault movement obviously being more recent than that. The Paunsaugunt Fault, along the east side of the plateau, is partly in the park. On it, displacement also is about 2000 feet (600 meters).

Several lesser faults, with offsets ranging from a few inches to several feet, parallel or cut across these two main faults. Some of these are shallow thrust faults thought to be the result of collapse and spreading of the Marysvale volcanic field to the north.

Bryce Canyon's sculptured walls reveal the layered lake deposits of the Claron Formation. Here, hoodoos are beginning to form. Vertical mud stalactites (far right) develop as rain washes mud from the hoodoos.
Ray Strauss photo.

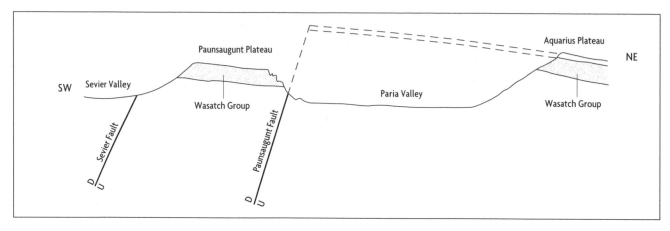

The Paria Amphitheater was carved in the uplifted block of the Aquarius Plateau. More Pink Cliffs appear along the edge of the higher plateau, raised well above the level of their counterparts at Bryce.

At Cedar Breaks, the Markagunt Plateau drops off abruptly along the Hurricane Fault, which effectively separates the Colorado Plateau as a whole from the Basin and Range region to the southwest.

At both Bryce and Cedar Breaks, many vertical joints and small-scale faults slice through the rock as well, most of them due to wrenching and breaking during uplift of the plateaus. Both faults and joints play important roles in directing erosion of the two dramatic escarpments.

For erosion is king here, erosion governed by plateau uplift, by downward erosion of the Colorado River and its tributaries, by relative weakness and porosity of the limestone layers, by joints and faults, by weather-

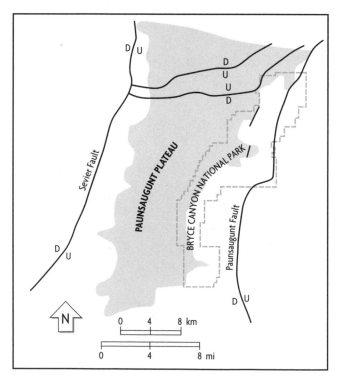

Two major faults define the Paunsaugunt Plateau. On both, the east side is lifted relative to the west side.

ing, and by climate. And what erosion! Towers and turrets, narrow scalloped walls, and thousands of ornate and fanciful figures populate the descending ramparts, delicately touched with color that varies in intensity from place to place as well as from hour to hour, day to day, and season to season.

The primary agent of this erosion is water, in the form of rain, frost, snowmelt, and runoff. The work of water is abetted by wind, frost, and gravity. Erosion initially followed joints and faults, which allowed passage of groundwater and solution of the limy calcium carbonate cement binding the siltstone and sandstone particles together. Gradually, joints and faults have widened into gullies.

As erosion claws at the edges of the Paunsaugunt and Markagunt Plateaus, the sculptured cliffs gradually retreat. In steep gulches, runoff from summer thunderstorms and spring snowmelt sweeps loose debris downward. Debris sliding through large gullies concentrates along their edges, where bits of rock and sand fall from the gully walls. As they move—either washed by water or sliding landslide fashion—they leave grooves or scour marks in the rock walls, showing past positions of the steep gully floors. Undermining resistant rock, their movement helps to maintain the steepness of gully walls.

Especially at Cedar Breaks, which is higher and colder than Bryce, water freezing in cracks and crevices pries rock blocks away from the upper parts of the cliffs. In the southern part of Bryce, where there are several prominent limestone and sandstone ledges, similar frost wedging loosens large rock blocks, which tumble from the cliff faces to slopes below. In both park areas, frost also works at a lesser scale, separating smaller blocks and even individual sand and silt particles from the porous, water-absorbing, loosely cemented rocks of the Claron Formation. As broken rock material falls, talus slopes and cones form below the cliffs; then frost heaving takes over, moving the talus fragments gradually downslope. Both frost wedging and frost heaving depend on day-night temperature variations, which are most extreme (with overnight freezing and daytime thawing some 200 to 300 times per year) on south-facing slopes near the rim.

Horizontal corrugations in the walls of sculptured alleyways and in the cliffs and turrets are due to natural variations in the composition of the sediments. Limestone and sandstone in general are more resistant than layers containing large proportions of silt and clay. Dolomite—like limestone but with added magnesium carbonate—is more resistant still. Dolomite and well-cemented sandstone cap much of the rim and many prominent ledges, walls, and hoodoos. In many places, the clayey residue from solution of calcium carbonate coats walls and hoodoos like stucco, concealing variation between the layers.

None of the erosion would be possible without differences in elevation caused by faulting. Normally, uplift sets the stage for erosion, as can be seen clearly at Cedar Breaks. But oddly, the Bryce portion of the landscape is on the low side of the controlling Paunsaugunt Fault. Land east of the fault, once raised higher, has been eroded even more severely, until the pink limestones and siltstones of the Claron Formation are gone entirely. Northeast of Bryce, where they

are protected by lava caprock, they still remain as the Pink Cliffs of the Aquarius Plateau.

Bryce and Cedar Breaks amphitheaters are still eroding headward, at a rate that can be determined by tree-ring studies. Many of the trees along the canyon rims now stand on tiptoe, their roots bared by continuing erosion of the edge of the canyon rim. When compared with a tree-ring "yardstick" developed by archaeologists for dating Southwest ruins, slender cores taken from these trees (with no harm to the trees) show that different parts of the rims recede 9 to 48 inches (24 to 120 centimeters) per century, a rapid change as geologic processes go. As the cliffs recede, small changes take place in the varied forms and figures of both amphitheaters. Some walls, some hoodoos fall; others, in unending succession, take their place.

Bryce Canyon National Park is on the ridgelike plateau between drainages of the Paria and Sevier Rivers. In dramatic contrast with its eastern slope, the western edge of the Paunsaugunt Plateau shows few of the intricately carved cliffs and pinnacles displayed in Bryce. There, summer temperatures are lower, snow insulates the ground in winter, moisture retention is higher, and vegetation is dense. Under these conditions erosion is much less severe. The Paria River and its

Hoodoos on the right skyline are remnants of a wall like that in the center of the photograph. Hoodoo development is most marked in the middle of the Claron Formation. Ray Strauss photo.

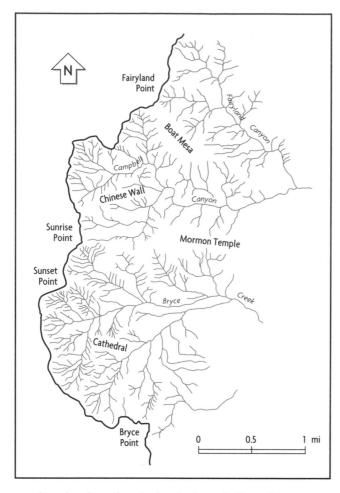

Eroding headward, Fairyland, Campbell, and Bryce Creeks have branched and rebranched in dendritic drainage patterns that locally, where gullies are parallel, reflect joint sets in the rocks.

branching tributaries erode the east side of the Paunsaugunt Plateau faster than the Sevier River and its tributaries erode the west side. Eventually the Paria tributaries will do away with the plateau altogether, breaking through the ridge and "capturing" the Sevier River's drainage.

GEOLOGIC HISTORY

Mesozoic and Cenozoic Eras. Paleozoic and early Mesozoic history are not represented in the Bryce and Cedar Breaks areas, but are similar to that of the rest of the Plateau. Thick layers of Cretaceous marine and near-shore sedimentary rocks underlie the Claron Formation. As North America drew away from Europe at the end of Mesozoic time, the continent rose slightly, and the widespread sea in which the rocks were deposited withdrew. In the continental interior two ranges rose destined to play parts in the Bryce–Cedar Breaks story: the Rocky Mountains in Wyoming and Colorado, and the Sevier Mountains in Utah. In landlocked areas between the two ranges, several large inland lakes developed, one of them stretching diagonally across Utah from northeast to southwest and measuring at its greatest extent about 75 by 250 miles (120 by 400 kilometers)—slightly larger than Lake Erie.

During Paleocene and Eocene time, as surrounding mountains eroded and their fine debris washed into intervening basins, the silty limestone and mud of the Claron Formation accumulated in the lakes and on floodplains of tributary rivers and streams.

Starting about 35 million years ago, in Oligocene time, and continuing well into Miocene time, volcanoes erupted nearby, creating, among other features, the lava flows that can now be seen atop the Aquarius Plateau. The Colorado Plateau as a whole began to rise, with volcanism and intrusions marking its margins. By late Miocene time the various segments of the Colorado Plateau had separated and moved up or down by different amounts.

Uplift was of course accompanied by erosion. When the ancestral Colorado River found a short route to sea level (see sidebar), headward erosion along some of the tributaries fairly rapidly (geologically speaking) created the Bryce Canyon and Cedar Breaks we see today.

BEHIND THE SCENES

Cedar Breaks. Headward erosion by Ashdown and Rattle Creeks and their small tributaries is responsible

REMOTE-CONTROL RIVER

Geologists now believe that 5.5 million years ago, quite recently geologically, the Colorado River found its way to the newly opened Gulf of California. Regenerated by the rapid drop to sea level, the river eroded powerfully, deepening the gorge we know as the Grand Canyon.

What does this have to do with Bryce and Cedar Breaks? For millions of years, as the Colorado carved deeper and deeper, its tributaries—among them the Paria and Virgin Rivers and their tributaries—also increased in power and deepened their canyons. Eroding headward, their headwaters encountered the relatively young, soft rocks of the Claron Formation, where erosion was particularly easy. There the smaller tributaries divided and redivided, etching smaller and smaller gullies, thus shaping the great semicircular amphitheaters of Bryce and Cedar Breaks.

Many other Plateau country features—the canyons, the lines of cliffs, the desert badlands—similarly owe their characteristics to the "remote control" of the Colorado River, and ultimately to the base level of the Gulf of California.

for this gigantic amphitheater in the edge of the Markagunt Plateau. The colorful, intricately carved cliffs have the same fairyland appearance as Bryce Canyon.

The Claron Formation is considerably thicker and more colorful here than at Bryce. We are near one of its main sources: the Sevier Mountains of early Tertiary time. The formation is also considerably sandier and siltier, and contains less limestone and dolomite and more volcanic ash, than its counterpart in Bryce. Grayish rocks underlying the pink strata are Cretaceous; many of them belong to the Kaiparowits Formation, made up of deposits of older rivers and freshwater or brackish lakes.

North of Cedar Breaks, Brian Head, elevation 11,307 feet (3466 meters), is capped with lava flows and volcanic ash that erupted in Eocene time. Explosive volcanic activity continued here throughout Tertiary time, with calderas and ash flows dominating the scenery. Quaternary lava flows, some of them quite fresh in appearance, surface parts of the Markagunt Plateau, in places damming small ponds and lakes.

There are quite a few bristlecone pines at Cedar Breaks; see the comment under Bryce Point, below.

Bryce Canyon viewpoints and trails are described below in north-to-south order, as they appear on entering the park by car.

Fairyland Viewpoint. The northernmost viewpoint in the park reveals the Pink Cliffs, carved in lake sediments of the Claron Formation, and provides a good view of lava-capped Aquarius Plateau to the north, an upfaulted relative of the Bryce area. Though this part of Bryce is not as deeply eroded as portions farther south, and slopes are not as steep, several interesting features are apparent. The Fairyland Fault transects Boat Mesa just to the south, offsetting some of the pink limestone and siltstone layers. Since the rimrock is not offset, movement along this fault must have preceded its deposition. On the Sinking Ship to the southeast, the Paunsaugunt Fault brings graybrown Cretaceous rocks hard up against pink limestones of the Claron Formation.

Fairyland Trail (8 miles or 13 kilometers). Going

Bands of hoodoos and finlike walls reflect the systematic layering of the lake deposits. Some layers erode into hoodoos; others form ledges or slopes. Ray Strauss photo.

down Fairyland Canyon, winding around the base of Boat Mesa, this trail offers good views of the tilted fault block of the Sinking Ship. The Paunsaugunt Fault cuts across the low "stern" of the ship, with the pink Claron lakebeds, tilted by drag along the fault, abutting gray and tan sandstone of the Wahweap and Straight Cliffs Formations.

A short side trail visits Tower Bridge. The main trail then climbs toward the Chinese Wall and returns to the rim near Sunrise Point. Good examples of differential erosion of hard and soft rock, and of the gullying process in general, appear along this part of the route, along with many views of other scenic and geologic features.

Sunrise Point. This viewpoint is on the dividing line between the less sharply eroded amphitheater formed by Campbell Creek and its tributaries and, south of it, the steeper, more ornate, more spectacular amphitheater of Bryce Creek and its tributaries. Several caprock layers control erosion here, with deep gully development initiated by erosion along joints.

Queens Garden Trail (1.5 miles or 2.4 kilometers) descends from Sunrise Point into an exceptionally sharply dissected part of the Bryce amphitheater, with steep cliffs and many pinnacles. Headward erosion of tributaries of Bryce Creek is at its most spectacular here. Gradients are very steep, and gully walls, protected by hard caprock and scoured by downward movement of debris in the gullies, are essentially vertical.

Sunset Point. Looking northeastward across the Paria Amphitheater to the Aquarius Plateau, southeastward toward Bryce Point, and downward into both the Queens Garden and the intersecting alleyways of the Silent City, views from Sunset Point are the most dramatic in the park. Resistant rock layers stand out above less resistant ones. Gullying is rapid, with headward erosion cutting back the rimrock and steep slopes below it, a process still going on. The amphitheater shape of the Silent City, formed by headward erosion of gullies tributary to Bryce Creek, shows up well between here and Bryce Point.

Inspiration Point provides another look down into the Silent City with its steep alleyways and light-reflecting walls. Between here and Bryce Point many

The Sinking Ship is a tilted fault block. Pink cliffs of the Aquarius Plateau show on the skyline. Ray Strauss photo.

large blocks of rimrock have fallen from the rim, pried loose along joints by frost wedging or undermined by erosion of softer rocks below.

Bryce Point. This high point is on an upraised block between two faults: the Bryce Point Fault (a reverse fault) and the Peekaboo Fault. The gnarled trees on Bryce Point (and also on Inspiration Point, along the rim of the Queens Garden, and near Rainbow and Yovimpa Points) are ancient bristlecone pines, the species that in California includes the oldest living things in the world, more than 4000 years old.

Peekaboo Loop Trail (3.5 miles or 5.5 kilometers). Dropping from Bryce Point into the Bryce Amphitheater, this trail covers some of the most dramatic ground in the park. If transportation is available, take this trail to its junction with the Queens Garden Trail, then ascend to the rim at Sunset Point. Along the trail are abundant examples of Bryce's erosional features.

Paria View. Looking out over the Paria Amphitheater, this view shows especially well some gently sloping mesas in the distance. These mesas are remnants of the floor of an earlier Paria Amphitheater, dissected probably less than 500,000 years ago during a renewed cycle of downward and headward erosion. Other features visible from this overlook include foreground foothills shaped in gray Cretaceous sandstone and shale, the lava-capped Aquarius Plateau to the northwest, and in clear weather the distant Henry Mountains, an eroded laccolith.

Natural Bridge. Formed where erosion by wind and rain attacked both sides of a narrow limestone fin or wall, this hole in the rock is a natural arch, not a true natural bridge, which would involve stream erosion. Note that it formed along a joint, where percolating water weakened the rock by dissolving its calcium carbonate cement, and freezing and thawing loosened its sand grains.

Rainbow and Yovimpa Points. At the southern end of the park road, these two points are unsurpassed for views to the south and southeast. The wide dome of the Kaibab Plateau, north of Grand Canyon, can be seen from Yovimpa Point. At both these viewpoints, erosion takes on a different character from that displayed farther north. Blocks of rimrock, wedged apart by frost and undermined by erosion of the soft sedimentary rocks below, have broken away and fallen. Accumulating on the talus apron below the cliffs, they move gradually downslope because of frost heaving. Frost action is a particularly potent factor here in the highest part of the park. Gray foothills below Yovimpa Point show that erosion reaches well down into Cretaceous rocks.

OTHER READING

Bezy, John, 1988. *Bryce Canyon: The Story Behind the Scenery*. KC Publications.

DeCourten, Frank, 1994. *Shadows of Time: The Geology of Bryce Canyon National Park*. Bryce Canyon Natural History Association.

Gregory, H. E., 1951. *The Geology and Geography of the Paunsaugunt Region, Utah*. U.S. Geological Survey Professional Paper 226.

CANYON DE CHELLY NATIONAL MONUMENT

ESTABLISHED: 1931
SIZE: 130 square miles (337 square kilometers)
ELEVATION: 5540 feet (1689 meters) at visitor center
ADDRESS: PO Box 588, Chinle, Arizona 86503-0588

STAR FEATURES

- ✪ Two scenic canyons with soaring walls of pink sandstone.
- ✪ A major anticline, the Defiance Uplift, that brings these rocks to our view.
- ✪ Natural rock alcoves sheltering homes of the prehistoric people who once lived in these canyons.
- ✪ Two mountain-fed streams that water farms of present-day Navajos.
- ✪ Rim drives, trails, canyon tours (with Navajo guides), visitor center/museum, and evening programs.

SETTING THE STAGE

The Defiance Uplift, an oval dome 100 miles (150 kilometers) long and 30 to 40 miles (50 to 65 kilometers) wide, is steeply inclined on its eastern flank but slopes more gently on its western side, where this national monument is located. Six thousand feet (2000 meters) higher than most of the Mesozoic sedimentary strata that surround it, the uplift has eroded down and into older rock layers, most notably the Permian De Chelly Sandstone, the pink sandstone exposed in the towering walls of Canyon del Muerto and Canyon de Chelly, both within the national monument.

Here the Defiance Uplift is surfaced with hard, resistant Shinarump Conglomerate, the lowest unit of the Triassic Chinle Formation. Lying directly on the De Chelly Sandstone, this resistant unit is a mixture of coarse sand and rounded pebbles of Precambrian rock and of particularly resistant rock such as chert, which is abundant as nodules in Paleozoic limestone. Short, straight, stream-type cross-bedding in the Shinarump Conglomerate shows up well when seen at a distance across the canyon from rim viewpoints.

The exposed upper surface of the Shinarump Conglomerate, scoured by wind and rain, is dotted with shallow potholes. After rains, when water stands in the potholes, tiny plants and animals populate these mini-ponds, living out their lives in the short days before their pools dry up. The little organisms secrete weak acids, products of their metabolism, that join with the

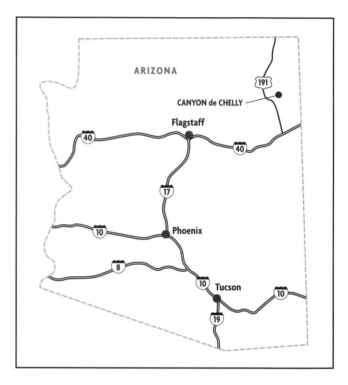

slight natural acidity of rainwater to attack the calcium carbonate cement that holds Shinarump sand grains together. When the ponds evaporate, wind blows away the loosened grains, deepening the potholes. Wind winnows other parts of the Shinarump surface, too, carrying away dust and sand but leaving the heavier pebbles as close-packed desert pavement.

From many of the rim viewpoints, the sharp contact between the Shinarump Conglomerate and the pale, peach-colored De Chelly Sandstone below it is clearly exposed, a contact that represents 60 million years of erosion or nondeposition. Here and there along the rim, stream channels filled with cross-bedded sandstone and conglomerate cut down into the De Chelly Sandstone.

Cross-bedding marks the De Chelly Sandstone with sweeping laminae and tells us of its wind-deposited origin. The sand grains are uniform in size, well rounded, and lightly frosted or pitted from collisions with other wind-borne sand grains. On the cliff walls, horizontal laminae intercept or bevel the sloping ones, marking transient interdune surfaces like those known amid modern dunes. On some of these flat surfaces, thin, flat-lying

The freestanding pinnacle of Spider Woman Rock (right) is an erosional remnant of De Chelly Sandstone remarkably similar to the not-yet-isolated rock to the left. Ray Strauss photo.

layers of silt accumulated, interdune deposits that appear and disappear over relatively short distances.

Undermined by stream erosion, this massive rock weakens along vertical joints, and ultimately giant slabs crash to the floor of the canyon. Undermining was doubtless much more forceful in the past, when Ice Age climates brought lower temperatures and increases in precipitation. The long, frigid winters of the Ice Ages brought to this area increased day-to-night and season-to-season temperature changes that repeatedly froze water in the joints, helping to pry slabs away from the walls.

The cross-bedding forms the sloping ceilings of many recesses and alcoves in Canyon de Chelly and Canyon del Muerto, particularly in the deeper eastern parts of the canyons. Localized concentration of moisture above silty, clayey, less permeable interdune deposits accounts for the erosion of most of these caves. As frost worked on rock already weakened by solution of its calcium carbonate cementing material, thin slabs and sheets of rock, and occasionally large blocks, fell away and deepened some of the recesses until they were large enough to be used as living space by Anasazi peoples who inhabited this region 700 to 1500 years ago.

The walls of Canyon de Chelly and Canyon del Muerto are streaked with shiny stains of manganese and iron oxides—desert varnish—and with dull ribbons of carbonaceous plant material washed from overlying soils or formed in place as algae, lichens, and mosses grew on frequently dampened surfaces. The smooth desert-varnish veneer of some rock faces tempted early inhabitants, as well as later Native Americans, to peck petroglyphs (rock pictures) through the dark sheen and into the pale rock behind.

The massive walls of Canyon de Chelly and Canyon del Muerto become higher and higher eastward as the De Chelly Sandstone rises toward the summit of the Defiance Plateau.

Hunters on horseback postdate Spanish introduction of horses. These petroglyphs were pecked through a layer of desert varnish. Halka Chronic photo.

GEOLOGIC HISTORY

Paleozoic Era. The Defiance Uplift seems to be a very old structure, one that more than once rose above the sea. On it, Permian rocks rest right on Precambrian rocks. It seems likely that most of the older Paleozoic rocks that occur elsewhere in the Plateau region were deposited here, but that they eroded off as the area lifted above sea level in Devonian and Pennsylvanian time. During Permian time, though the area still tended to remain high relative to its surroundings, deposition reached across the arch of the uplift again. The deposits—some marine, some continental—demonstrate the nearness of the Permian shore.

Cross-beds in the De Chelly Sandstone almost all slope southwestward, showing that prevailing winds that formed them blew predominantly from today's northeast. The greatest present-day deserts, the Sahara and Arabian Deserts and the dry interior of Australia, lie 20 to 30 degrees north and south of the equator, in zones where easterly trade winds of the tropics meet westerlies of temperate belts. The deserts of Permian time, which extended north and east beyond Arizona's boundaries, lay in a similar belt relative to the position of the Permian equator, which ran obliquely across what is now North America.

Mesozoic Era. In early Triassic time, mountains rose in central and southern Arizona, high ranges that shed coarse sand and gravel northward across a broad, flat coastal plain. A thin layer of debris from these mountains is today the Shinarump Conglomerate, hard rimrock of Canyon de Chelly and Canyon del Muerto. Older highlands in northwestern New Mexico and southwestern Colorado may also have supplied rock debris to this area.

In most of northeastern Arizona the Shinarump Conglomerate is underlain by Triassic red-brown shales and mudstone—the "redbeds" of the Moenkopi Formation. Here on the Defiance Uplift, however, no such redbeds exist. Their absence, as well as the absence of many Paleozoic strata that we might expect below them, is evidence that the Defiance Uplift has been present, off and on, for hundreds of millions of years.

Later in Triassic time the nature of the deposits changed. Volcanoes somewhere to windward contributed great quantities of volcanic ash, the raw material of most of the Chinle Formation. Decomposing into several different clay minerals of a group known collectively as bentonite, the volcanic ash today is responsible for the many-hued badlands of the Painted Desert (see sidebar on page 132). Some of it seems to have been stream deposited, probably carried by floods far greater than but similar to those that followed the 1980 Mount St. Helens eruption.

Late Triassic and early Jurassic time saw the accumulation of great seas of windblown sand across a broad desert that must have resembled the Sahara and Arabian Deserts of today. Many hundreds of feet of dune

Towering walls of De Chelly Sandstone reveal sloping cross-bedding of Permian sand dunes. Carbonaceous plant material derived from lichens adds long streaks above White House Ruin. Halka Chronic photo.

sand—now the Navajo Sandstone—accumulated in deserts that stretched from southern Nevada to Wyoming. Then, after several rounds of river and stream and volcanic ash deposits, the land was once more beveled by erosion.

During Cretaceous time a wide sag developed across a region that extended far north and east of Arizona. An arm of the Cretaceous interior sea crept in across this part of the continent, and as it retreated it deposited a thin layer of near-shore sands—the Dakota Sandstone. A second and longer-lasting advance left a thick layer of dark gray marine shale—the Mancos Shale. As the sea retreated, these deposits were capped by near-shore, lagoon, and beach deposits, the sandstone, shale, and coal sequence of the Mesaverde Group.

None of these Jurassic and Cretaceous rocks appear in Canyon de Chelly National Monument, but they can be seen across Chinle Valley in the sloping walls and cliff-edged cap of Black Mesa.

Cenozoic Era. Within the national monument no record remains for the Cenozoic Era. But not far to the south, lake- and stream-deposited sandstone, conglomerate, and limestone layers of the Bidahochi Formation tell of a large Miocene-Pliocene lake that along its northeastern edge slowly filled with sand and gravel washed off the Defiance Uplift. The Chuska Sandstone, high up on the Chuska Mountains east of Canyon de Chelly, is of about the same age; its lower part was clearly deposited by streams, and its upper part as dunes.

Volcanism played a role in establishing the Chuska Mountains: Lava flows and fragmented and rewelded volcanic breccia cap this range and shield it from the erosion that has bared much of the surrounding country. Though the Chuskas were untouched by glaciation, the heavier rainfall and snowfall of Pleistocene time did facilitate slumps and landslides in these mountains and doubtless affected rates of downcutting in Canyon del Muerto and Canyon de Chelly.

OTHER READING

Anderson, Barbara, et al., 1990. *Canyon de Chelly: The Story Behind the Scenery.* KC Publications.
Grant, Campbell, 1978. *Canyon de Chelly: Its People and Rock Art.* University of Arizona Press.

CANYONLANDS NATIONAL PARK

ESTABLISHED: 1964
SIZE: 527 square miles (1367 square kilometers)
ELEVATION: 3680 to 6560 feet (1125 to 2006 meters)
ADDRESS: 2282 S. West Blvd., Moab, Utah 84532-3298

STAR FEATURES

✪ Well-exposed "layer cake" geology with vivid erosion-carved tiers of Paleozoic and Mesozoic sedimentary rocks.

✪ A dramatic landscape of deep canyons separating mesas, buttes, and pinnacles shaped by water, wind, and gravity.

✪ Two great rivers, masters of canyon country erosion.

✪ Anticlines, grabens, and other geologic structures created by slow-flowing subterranean salt.

✪ An unusual and interesting dome thought to be the result of meteorite impact or doming salt.

✪ Visitor center, guided tours, trail and roadside exhibits, guide leaflets, and evening programs.

See color pages for additional photographs.

SETTING THE STAGE

Where two mighty rivers meet, they have shaped a lonesome, dramatic land of barren canyons, cliffs, buttes, and mesas, a strangely beautiful, awesome region where erosion rules. This is harsh country, dissected by tortuous canyons that are obstacles to travel and exploration. Yet it is fascinating country, because here the geology is written boldly and clearly on the denuded landscape.

Sedimentary rocks in this national park, flat-lying or arching very gently across the anticline of the broad Monument Uplift and locally bowed upward by the movement of underlying salt, have eroded into layered scenery where alternating bands of resistant sandstone and less resistant siltstone form a ledge-slope-ledge topography. Above particularly resistant cliff-formers, slopes flatten into nearly horizontal benches, natural landings between the plateau surface and the rivers.

Many cliff-forming sandstones show the broad, sloping cross-bedding and fine, rounded and frosted grains that characterize dune sand. Siltstone and mudstone slope-formers, deposited in ancient deltas or on wide river floodplains, still bear ripple marks and raindrop impressions formed before they hardened into

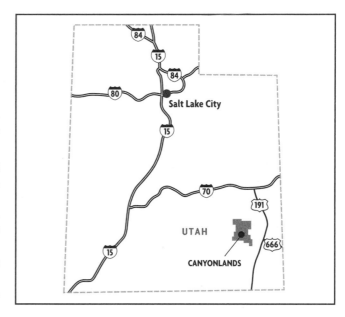

rock. Fossils are extremely rare except in a few thin limestone layers in the lowest part of the rock sequence, and some dinosaur tracks in Mesozoic rocks.

Almost all of these rocks are pink, brick red, or salmon-colored, tinted with varying amounts of hematite. Some rock surfaces are coated with dark brown or blue-black desert varnish (see sidebar).

In regions of flat-lying, uniform strata, streams normally develop dendritic drainage patterns, branching like trees. Here this pattern is carried to extremes. Tributary stream courses divide and redivide until they are the fine rills that decorate the highest slopes of the canyon walls. Wind, gravity, and short-lived torrents fed by sudden storms wash and wear away siltstone and mudstone, and undermine sandstone cliffs so that great rock slabs fall away along vertical joints. Century by century, canyons widen and plateaus reduce to mesas, mesas to buttes, buttes to pinnacles, until nothing is left but piles of rocky rubble. Eventually the rubble breaks up too, and is blown away or carried by storm-born streams to the two great rivers that flow through the park.

The Colorado and Green Rivers, with headwaters in the Rocky Mountains of Colorado and Wyoming, regulate the depth of side canyons, for tributaries can't erode below the level where they join their main stream. Placid as they enter the park, the two rivers join forces

GLOSSING IT OVER

Desert varnish, the smooth, sometimes iridescent film that coats many rock faces in the Plateau country, is a thin layer of iron and manganese oxides and silica. Unfortunately, the term is sometimes also used for the much duller surfaces of lichens and mosses that mark cliffs with long, V-shaped streaks.

The origin of desert varnish is somewhat of a mystery. For a time geologists thought its minerals were carried from the rock by evaporation of moisture within the rock, or by the surface weathering of rock by lichens and other small organisms. But analysis of the underlying rocks revealed differences in composition not reflected in the composition of the desert varnish. In fact, some "varnished" rock didn't contain the necessary elements at all.

So where did it come from? Because it forms on rocks of differing compositions, it must come from an external source. The only widespread external source seems to be dust blown in from other areas. When the dust reaches a rock surface (perhaps brought down by rain), iron and manganese and silica from the dust may be fixed to the rock by microorganisms; other elements are washed off or blown away. Because the elements are processed in only minute amounts, and perhaps at infrequent intervals, the dark coating on the rocks forms extremely slowly, over thousands of years. Desert heat seems to be a contributing factor.

Both before and after the coming of Europeans, native peoples in the Plateau area have pecked through the dark desert varnish to inscribe designs and perhaps messages into the light-colored rock below, as have desert dwellers in many other parts of the world.

A thick layer of desert varnish over light-colored sandstone makes an ideal scratchboard in this bright display of petroglyphs. These are at Newspaper Rock just outside of Canyonlands National Park. Halka Chronic photo.

Permian rocks of the Needles area, their sand, silt, and pebbles washed from Uncompahgria in Permian time, thicken, thin, or disappear over short horizontal distances. Ray Strauss photo.

to rampage through Cataract Canyon, one of the wildest stretches of water in any national park, its lower end now stilled by Lake Powell. Furthermore, the two rivers follow twisting courses that reveal oddities of their earlier history, when they meandered over a wide plain covered with Tertiary sediments.

Along the canyons of the Green and Colorado, strata bow up slightly toward each river, partly because of the gentle, almost imperceptible arch of the Monument Uplift, and partly in response to the upward push of thick layers of salt some distance below the surface. Salt becomes plastic under pressure, flowing extremely slowly, moving from regions where pressure is great to regions of lower pressure. And because it is lighter than most rock, it tends to push up wherever it can.

Pressure on the salt is mostly due to overburden, the weight of thousands of feet of overlying strata, including a great thickness of Tertiary sediments now eroded away completely. Where the Green and Colorado Rivers and their larger tributaries removed some of this overburden, the pressure on the salt lessened. And so through millions of years the salt slowly flowed toward the rivers, bowing up the strata so gradually that the rivers never fluctuated from their courses. An alternate explanation of the salt anticlines along the rivers places them directly over large faults in Precambrian rocks, as at Arches National Park. The salt in this case flowed toward the fault ridges, forming anticlines above them, and the rivers later established courses along valleys where the salt had dissolved and overlying rocks had collapsed. It is possible these two theoretical mechanisms worked together, each augmenting the other.

GEOLOGIC HISTORY

Paleozoic Era. For much of Paleozoic time this area lay beneath a shallow sea, where successive layers of marine limestone, sandstone, and shale were deposited. None of the older marine strata appear in the park, but thick beds of Pennsylvanian salt influence the present park picture, so let's start our discussion with their deposition.

In Pennsylvanian time, the ancestral Rocky Mountains rose along preexisting fault lines in what is now Colorado. West of Uncompahgria, the westernmost of these ranges, the faulting also created a broad basin, 12,000 feet (4000 meters) deep and almost cut off from the sea—a veritable Mediterranean. Now called the Paradox Basin, it was at least periodically flooded with sea water. As is happening in the Mediterranean Sea today, the almost isolated body of water became briny, and salt, potash, anhydrite, and gypsum were deposited, layer by layer, in its depths, with a total thickness

of close to an astonishing 6000 feet (1800 meters). Twenty-nine layers, separated by thin black and brown shales that may indicate either stagnant conditions or new influxes of sea water, have been recognized in oil well cores from this region. Worldwide fluctuations in sea level, showing up as cyclic deposits—alternating sandstone-shale or shale-limestone sequences—are known in Pennsylvanian strata in many parts of the world, and may have been responsible for these cyclic salt deposits, which together make up the Paradox Salt, or the Paradox Formation.

Another 6000 feet (1800 meters) of Permian sediment eventually filled the Paradox Basin: Reddish conglomerate, sandstone, and siltstone of the Cutler Formation, washed from Uncompahgria, intertongues with light-colored beach and dune sands of the Cedar Mesa Sandstone. The red color of the Cutler Formation is due to small amounts of the mineral hematite, iron oxide derived from the dark minerals coming from Uncompahgria.

Mesozoic Era. In Triassic time, rivers and streams continued to bring mud, silt, and sand from the uplifts in Colorado and New Mexico, and deposited them in broad, thin sheets on floodplains, tidal flats, and deltas, with rivers flowing northwest to a continental seaway. The dark red Moenkopi Formation, its color

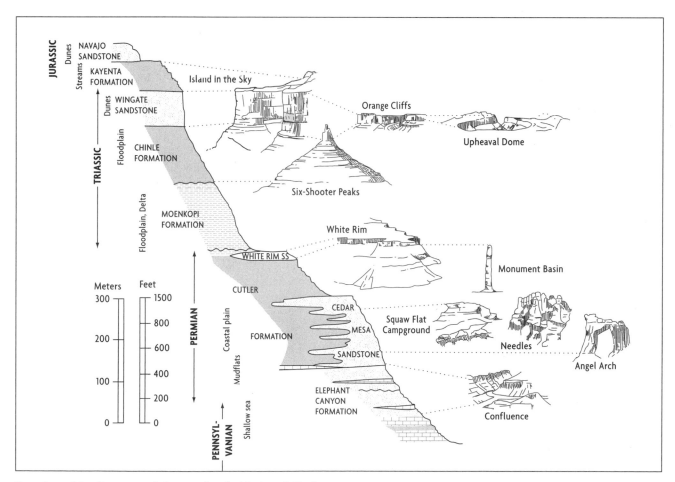

Stratigraphic diagram of Canyonlands National Park

In the Needles District, where sets of parallel joints intersect at almost right angles, the interlayered Permian Cedar Mesa Sandstone and Cutler Formation erode into a forest of fantastic pinnacles. Ray Strauss photo.

also derived from dark minerals coming from Uncompahgria, contains abundant evidence of tidal-flat deposition in its numerous small-scale ripple marks, raindrop impressions, burrows, and mud cracks. Above it the Chinle Formation contains silt and clay heavily infused with volcanic ash.

After an interval of erosion or nondeposition, Jurassic dunes swept across this area, leaving two prominent, easily recognized layers of eolian sandstone, distinctly cross-bedded and made up of even-sized, rounded and frosted sand grains. Reddish siltstone bands within them represent interdune deposits. The lower of the eolian formations, the Wingate Sandstone, appears as steep reddish-brown cliffs, commonly darkened with desert varnish, that surround the southern part of the monument. It is overlain by the Kayenta Formation, reddish sands and muds deposited on river floodplains where dinosaurs wandered—and left their tracks. The Navajo Sandstone, the upper of the two eolian sandstones, lies above the Kayenta "redbeds." It is lighter in color than the Wingate, and often erodes into rounded mounds on which sweeping diagonal dune-style cross-beds, etched by wind and rain, are easily seen.

The sea—this time coming from the east—invaded this area again in Cretaceous time. Thick marine shale and sandstone layers deposited in it have now been stripped away in the Canyonlands area, but they are known from such nearby areas as Mesa Verde National Park. Late in Cretaceous time the sea withdrew eastward as the land began to rise—the first hint of the coming Laramide Orogeny, when the building of the Rocky Mountains would completely change the character of this part of the continent.

Cenozoic Era. Although the park contains no Tertiary deposits, we know from adjacent regions that thousands of meters of silt, clay, and fine sand stripped from the newborn Rocky Mountains were deposited here in early Tertiary time. Many of these rocks contained glassy shards of volcanic ash, reflecting volcanic activity in surrounding areas.

Later in Tertiary time the whole Rocky Mountain region, including eastern Utah, bowed upward, rising as much as 5000 feet (1800 meters) above its former position. Rivers and streams, their courses steepened, scoured ever more deeply, washing away the fine, unconsolidated silt, sand, and volcanic ash deposited in early Tertiary time. The Colorado and Green Rivers eventually established courses across the broad plains west of the mountains, coming together in what is now Canyonlands National Park, their courses perhaps dictated by ancient fault zones and the rise of salt anticlines. In soft sediments, the rivers deepened their meandering channels. Soon imprisoned in steep-walled gorges, they nevertheless retained their twisting, winding courses. And with the help of thousands of small tributaries, most of them flowing only after heavy rains, these great rivers carved Canyonlands as we see it today.

BEHIND THE SCENES

Cataract Canyon. Viewed from the rim of the canyon above their confluence, the Colorado and Green Rivers flow smoothly, with gradients of about 1 foot per mile (0.2 meter per kilometer), gradually blending their waters. But several miles downstream the combined rivers—now called the Colorado—surge into Cataract Canyon, named by John Wesley Powell in 1869 as he led the first Colorado River float trip. There the Colorado plunges again and again through churning rapids caused by rock debris washed in from tributary canyons, by massive talus cones below vertical canyon walls, by huge fallen boulders and slumps, and by small landslides that constrict the stream. More than half of the rapids encountered by Powell are now drowned beneath the waters of Lake Powell.

Along parts of Cataract Canyon, rock strata dip away from the river because of the rise of buried salt. Rocks near the river are the oldest in the park: Permian marine limestone and, below it, fossil-bearing Pennsylvanian marine limestone, sandstone, and shale. Gypsum, less soluble than salt but similarly becoming plastic with pressure, and part of the Pennsylvanian evaporite deposits breaks through to the surface in the depths of Cataract Canyon near the mouths of Red Lake and Gypsum Canyons.

Colorado and Green Rivers. Above their confluence the Green and Colorado Rivers are docile and easygoing, essentially free from rapids. They swing back and forth in meanders inherited from Miocene time, when they looped across a nearly horizontal plain. The rivers have shortcut some of their entrenched meanders, leaving abandoned loops high above present river level. The most prominent of these is outside the park: Jackson Hole, across the Colorado River from Potash. Anderson Bottom, on the Green River, is another.

The Confluence. The junction of the Green and Colorado Rivers is often made interesting by differences in the amount of sediment they carry. Sometimes the Colorado is muddier than the Green; sometimes it

Druid Arch formed in a thin fin of Cedar Mesa Sandstone. Ray Strauss photo.

is clearer. For some distance after they join, the clear and muddy waters flow side by side, unmixed. Irregularities in the channel cause turbulence and eventual mixing, however.

In times of low water, sandbars and beaches develop here. Salt cedars (tamarisks) growing on the sand have stabilized many of these bars and beaches. Salt cedar is a newcomer, a native of the Near East, introduced into Mexico. It has now invaded most Southwest rivers, particularly flood-free ones controlled by dams, thriving wherever it is well fed by river or spring waters. Native plants can't compete with it, and its introduction reduces natural forage for wildlife. It is a prodigious water user, able to dry up natural springs.

The Grabens. Accessible only on foot or by four-wheel-drive vehicle, the Grabens are slender, flat-floored chasms as much as 300 feet (100 meters) deep. In the Needles fault zone, the Grabens occur where parallel vertical faults separate slim slices of rock. There are quite a number of these linear valleys here; four are traversed by four-wheel-drive trails: Devils Canyon, Devils Lane, Cyclone Canyon, and the unnamed graben leading to Beef Basin.

The origin of the Grabens lies with movement of subterranean salt out of this area and into the salt-cored anticline below the nearby Colorado River, and possibly with younger rocks gliding on a sloping salt surface, created as salt moved toward the anticline along the river. The Grabens are geologically young; many still have smooth, essentially vertical fault-scarp walls almost untouched by erosion. Cyclone Canyon and several smaller grabens have not yet developed through drainage.

Grand View Point. From this high, readily accessible vantage point, with magnificent views out over almost all of Canyonlands National Park, you can easily identify the rock layers by their expression as cliffs, ledges, slopes, and wide benches. White Kayenta Sandstone surfaces the point; brick red mudstone of the Elephant Canyon Formation appears near the rivers. Notice particularly the tall cliffs of Wingate Sandstone, dark with streaks of lichens and desert varnish, and the conspicuous wide bench of White Rim Sandstone at the top of the Cutler Formation. The White Rim Sandstone thickens westward, forming a higher and higher cliff below this bench. The tiny track winding along the bench is the White Rim Road, suitable only for four-wheel-drive vehicles. To the southwest the Green River negotiates a hairpin turn in one of its entrenched meanders.

In Monument Basin, southeast of the overlook and about 2 miles (3 kilometers) away, below the White Rim bench, a tall finger of Cutler Formation indicates the scale of the surrounding country. You can see about two-thirds of this rock totem pole, which is 305 feet (94 meters) tall and less than 38 feet (12 meters) thick.

Three mountain ranges break the skyline: the La Sal Mountains to the northeast, the Abajo Mountains to the southeast, and the Henry Mountains to the southwest. The three ranges formed as igneous rock welled upward in mid-Tertiary time, doming sedimentary layers but not, as far as we know, breaking through to the surface. On the horizon just west of the Abajo Mountains, a broad arch of sedimentary rocks extends south toward Arizona.

Between here and the Abajo Mountains, you can also distinguish the finely divided red spires of the Needles District of the national park.

Island in the Sky. Riding the wedge between the Green River on the west and the Colorado on the east, Island in the Sky rises 2000 feet (600 meters) above the two rivers, offering excellent views of the whole Canyonlands area. From this vista you can see an incredible maze of benches, canyons, buttes, pinnacles, and brilliant-hued rocks in an all-pervading layer-cake pattern.

Island in the Sky is a Y-shaped mesa standing almost free of the country to the north. You enter the Y along its eastern arm, crossing the slender "Neck" that is its only connection with the "mainland." Ultimately, as cliffs on either side are undermined by erosion, the Neck will fall, isolating the Island completely.

Near its edges the Island is surfaced with ledgy layers of Kayenta Sandstone; in its highest portions are rounded knolls and knobs of cross-bedded, white Navajo Sandstone, a thick dune sandstone that covers much of Utah, northeast Arizona, and western Colorado. High cliffs that drop off vertically all around the Island are the Wingate Formation. Lower layers are similar to those seen from Grand View Point.

The Maze, Land of Standing Rocks, and Elaterite Basin. The part of the national park west of the rivers can be reached only by four-wheel-drive vehicles. The Land of Standing Rocks is part of a broad bench on the Cedar Mesa Sandstone, part of the Cutler Formation. Stream patterns are dendritic in most of this area, but controlled by joints in such places as the Fins. Several rock units, including the White Rim Sandstone and parts of the Moenkopi Formation, thin or thicken between layers of siltstone. Such changes are common; they show up particularly well here because of contrasts in rock color. They represent horizontal changes in deposition, where the kind of sediment changes from place to place, as in beaches and muddy lagoons, dunes and interdunes, or sandy channels in silty floodplains. Watch for these features and for faults, which show up as offsets of the horizontal rock layers.

Elaterite Basin, mostly outside the national park, and Elaterite Butte, within the park, are named for a dark brown tarlike mineral that seeps from parts of the White Rim Sandstone. Elaterite originates from organic matter, as does its relative, oil, and may travel some distance from its source to lodge in a porous reservoir rock.

Island in the Sky looks down on a broad bench of White Rim Sandstone, above the scalloped margins of an entrenched meander of the Green River. Ray Strauss photo.

Needles District. The broad, level surfaces of the Needles District are part of a wide bench between the rim of the canyon country and the inner chasm of the Colorado River. The rocks that surface most of the bench are interlayers of the Cedar Mesa Sandstone, a white coastal sand unit, and the red layers of the Cutler Formation, originally eroded from the Ancestral Rocky Mountains. The striped appearance of this rock results from their interlayering.

The White Rim Sandstone, which makes a prominent bench in the northern part of Canyonlands, is not present here; bright vermilion siltstone and sandstone occupy its position at the top of the Cutler Formation. These rocks appear in the lowest, orangest slopes of North and South Sixshooter Peaks and at corresponding levels in the cliff facade to the east. Higher parts of these slopes are the Moenkopi and Chinle Formations. The cliffs themselves, darkened with desert varnish, are Wingate Sandstone.

The Needles District owes its carved turrets and occasional arches to weathering and erosion of Cedar Mesa Sandstone along two intersecting sets of joints and faults, one paralleling the Grabens, the other at almost right angles. In the enchanted world of the Needles themselves, joints are particularly closely and regularly spaced. Dune sandstone such as that of the Cedar Mesa Sandstone shows well-sorted and evenly rounded grains; its smooth texture lets it break away, where it is undermined, in caves and natural arches. All the arches in this part of the park are in this sandstone. In places, faults offset the red and white rock layers.

The Potholes. Where the undulating surface of the Cedar Mesa Sandstone collects rainwater, shallow pools nourish tiny plants and animals born of spores, eggs, and seeds that can survive long droughts between rains. Small shrubs and grasses grow nearby, their roots searching rock joints for precious moisture. Metabolic acids secreted by these organisms add to the slight natural acidity of rainwater, so that the water in pools and crevices dissolves the calcium carbonate cement that holds together the sand grains of the sandstone. Little by little, grain by grain, helped by wind and frost, slight depressions become deep potholes, more and more able to hold rainwater and so to perpetuate the cycle.

Salt Canyon and Horse Canyon. Four-wheel-drive trails threading these canyons, and foot trails branching from them, lead to Paul Bunyan's Potty, Gothic Arch, Castle Arch, and Angel Arch, as well as to prehistoric rock art and small ruins. (For a discussion on the origin of arches, see Arches National Park.)

Horse Canyon is relatively straight and shallow, sand floored, and walled only with Cedar Mesa Sandstone. Its drainage area is small, rarely providing enough water to clear the canyon down to bedrock.

Salt Canyon, on the other hand, is winding and rock floored. Its drainage area is larger, and the canyon as a result cuts through the Cedar Mesa Sandstone and into the underlying Elephant Canyon Formation, following a twisting course along joints related to the Needles fault zone.

Shafer Trail. This road, built for access to uranium mines, zigzags dizzily down the vertical bastion of Island in the Sky, dropping through almost all the rock units exposed in the park. Ledgy cliffs of Kayenta Sandstone, and below them a precipice of Wingate Sandstone, border the road's initial descent. Below the cliffs, the route provides a good look at the Moenkopi and Chinle Formations just above the White Rim. Some of the mudstones of these formations bear ripple marks and raindrop impressions formed 200 million years ago. In canyons and gullies below the White Rim are red and purple sandstone and conglomerate of the Cutler and Elephant Canyon Formations. A few thin limestone layers in the Elephant Canyon contain fossil shellfish of very late Pennsylvanian and early Permian age.

Upheaval Dome. This startling circular structure, a mound of Triassic sandstone and mudstone surrounded by concentric rings of much-faulted younger rocks, was long thought to be due to salt from the Paradox Formation slowly pushing upward toward the surface, as in the salt domes of Louisiana and Texas. Many geologists now think it is an impact structure, however, the result of a meteorite slamming into the Earth's surface here millions of years ago. Striking with enough force to penetrate hundreds of feet of then-overlying rock, the meteorite blasted a crater or perhaps an underground chamber several miles

With its seemingly structureless central mass and its concentric rings of Triassic and Jurassic rock, Upheaval Dome probably took shape at about the same time as the great extinction at the end of Cretaceous time. National Park Service photo.

across. The walls of the crater or underground cavity immediately collapsed, sliding inward along curving faults, converging to force up a central peak-shaped uplift. What we are seeing now is the lower part of the central peak, surrounded by some of the shattered remains of rocks from the crater or cavity walls.

Seismic studies—analyses of reflections of explosion shock waves bounced off the rocks below—show that the shattered zone extends downward to the top of the thick salt layers of the Paradox Formation. The top of the salt (actually salt, gypsum, and other evaporite minerals) is deformed upward by about 300 feet.

Study of the rocks distorted by the impact, and projection of the strata that once lay above them, suggest that the impact may have occurred in very late Cretaceous time or very early Tertiary time. No trace of a meteorite has ever been found.

White Rim Trail. This four-wheel-drive track follows a tortuous course from Potash on the Colorado River, south past the base of the Shafer Trail and Grand View Point, and then north to Horsethief Trail, outside the park on the Green River. For almost all of this distance the track rides the upper surface of the White Rim Sandstone at the top of the Cutler Formation. Strata visible along the route are the same as those described for Grand View Point.

OTHER READING

Baars, Donald L., 1994. *Canyonlands Country: Geology of Canyonlands and Arches National Parks.* University of Utah Press.

Childs, Craig Leland, 1995. *Stone Desert: A Naturalist's Exploration of Canyonlands National Park.* Westcliffe Publishers.

Huntoon, P. W., G. H. Billingsley Jr., and W. J. Breed, 1982. *Geologic Map of Canyonlands National Park and Vicinity, Utah.* Canyonlands Natural History Association.

Johnson, David M., 1997. *Canyonlands: The Story Behind the Scenery.* KC Publications.

CAPITOL REEF NATIONAL PARK

ESTABLISHED: 1937 as a national monument, 1971 as a national park
SIZE: 382 square miles (989 square kilometers)
ELEVATION: 5500 feet (1676 meters) at visitor center
ADDRESS: HC 70, Box 15, Torrey, Utah 84775-9602

STAR FEATURES

✪ A dramatic monocline in otherwise nearly flat-lying sedimentary rocks.

✪ Narrow, high-walled canyons, rock arches, natural bridges, and "waterpockets" eroded in dune-formed sandstone.

✪ A well-exposed sequence of color-splashed Mesozoic sedimentary rocks carrying records of floodplain, delta, tidal-flat, and desert environments.

✪ Indirect evidence of Pleistocene glaciation.

✪ Visitor center, museum, scenic drives and trails (some with guide leaflets), guided hikes, and evening programs. A relief map in the visitor center presents an excellent overview of the park and its geology.

See page 19 and color pages for additional photographs.

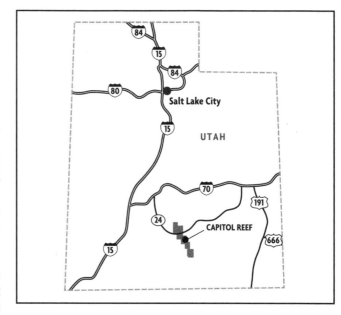

SETTING THE STAGE

The major scenic feature in this national park is a single prominent monocline known as Waterpocket Fold. Involving strata of Permian, Triassic, Jurassic, and Cretaceous age, steeply dipping toward the east, this fold is less than 3 miles (5 kilometers) wide but nearly 100 miles (160 kilometers) long. Along it, nearly horizontal rock layers suddenly plunge eastward like a breaking wave, and then flatten out again at a lower level. Because each rock type responds differently to erosion, a parallel array of ridges and valleys marks the position of the fold. The highest ridge, in the most massive and resistant rock layers, is known as Capitol Reef.

The name comes from several rounded bare-rock summits that resemble capitol domes, coupled with an old word for a rocky barrier (at sea or on land). And a barrier it is, with only four canyons (all of them subject to flash floods) allowing passage for water, animals, and people. The name of Waterpocket Fold stems from natural rain-filled pools, many of them pocket-shaped, in the massive sandstone at the heart of Capitol Reef.

One of the series of parallel monoclines that mark the Plateau country, Waterpocket Fold extends from Thousand Lake Mountain just north of the national park, south and southeastward to Lake Powell. Along its west side the fold is marked by a high, angular cliff of Wingate Sandstone, red tinted and daubed and streaked with dark lichens and desert varnish. The cliff rises sharply above slopes of soft green and purple shale and volcanic ash of the Chinle Formation, which in turn lies on dark red sandstone and siltstone of the Moenkopi Formation, the surface rock in western parts of the park. Erosion of the Moenkopi and Chinle Formations undermines the massive Wingate cliff, so that from time to time giant slabs, no longer supported at their base, break off along vertical joints and crash to the slope below.

Above and east of the cliff-forming Wingate Sandstone, separated from it by thinner layers of the Kayenta Formation, is another thick unit, the Navajo Sandstone. Not as well compacted or cemented as the Wingate, the Navajo Sandstone weathers into rounded cliffs and uplands—the barren "slickrock" of the summit of Waterpocket Fold. East of the summit, the denuded upper surface of this formation is fully exposed as the east flank of Capitol Reef. Thus the reef itself is asymmetrical: angular cliffs on the west, rounded bare-rock slopes on the east.

Colorful Triassic rocks, from the Moenkopi Formation (bottom) to the Wingate Sandstone (top), form the bold western face of Capitol Reef. Note the vertical joints in the Wingate Sandstone. Ray Strauss photo.

Both the Wingate and the Navajo Sandstone are marked with long, sweeping cross-bedding and have the fine, rounded and frosted grains of dune-formed sandstone. Wind-formed ripple marks decorate many rock surfaces. (Unlike ripple marks formed by water, which are always at right angles to the stream flow, dune ripples may run up and down dune surfaces.)

No in-place lava flows exist within the park, but near the Fremont River boulders of dark gray basalt veneer flat-topped hills. Well rounded by stream action, these boulders were obviously carried from basalt-capped plateaus to the west. In Cathedral Valley, in the northern part of the park, dikes and volcanic necks mark sites of Tertiary volcanoes probably older than the plateau basalts.

Though only four canyons penetrate Capitol Reef's entire thickness, the monocline is a labyrinth of canyons, some sheer-walled and sandy-floored. Important factors in their development are day-night temperature contrasts and the cloudburst climate of this part of the Southwest. Temperature contrasts that cause freezing and thawing of small amounts of moisture, either in joints or between sand grains of porous rock, are insignificant in the short term, but repeated again and again they eventually loosen sand grains and disintegrate exposed rock surfaces. Such frost wedging is especially effective in shady clefts and narrow defiles, where moisture is retained for longer periods.

Frost may initiate development of waterpockets, too, enlarging slight depressions created by wind erosion. As the depressions deepen, as they hold more water for longer periods of time, spores and seeds of small plants and larvae of tiny animals are brought to them by wind or by birds, lizards, and other animals coming to the pools to drink. Developing in the pools, the little plants and larvae, many of them microscopic, secrete acid metabolic products that dissolve the calcium carbonate cement of surrounding sandstone. Then, when the pools dry up, wind blows away the loose grains, deepening the hollows.

Small rock depressions may string together, particularly along joints, guiding the flow of rainwater, inaugurating development of clefts, crevasses, and eventually deep, narrow canyons. Water from sudden downpours churns through these passageways, scouring and smoothing their floors and walls. The swirling streams may also grind out large potholes or carve arches and natural bridges. (For more on arch formation, see Arches National Park.)

GEOLOGIC HISTORY

Sedimentary rocks here range from Permian sandstone and marine limestone in several deep canyons along the west edge of the park, to Cretaceous shale and sandstone in parallel valleys and ridges east of Capitol Reef. Representing a time span of about 200 million years, these rocks were deposited in a long succession of changing environments.

Paleozoic Era. The story of this park begins 260 million years ago on a Permian desert, with sand dunes that have become the Cedar Mesa Sandstone, now visible in the bottoms of the deeply entrenched canyons of Sulphur and Fremont Creeks in the western part of the park. Above the Cedar Mesa Sandstone are light gray ledges of Kaibab Limestone, the rimrock of the Grand Canyon, here sandy and impure because it was deposited close to the shore of the Permian sea, the last Paleozoic sea to transgress the continent.

Mesozoic Era. In Triassic time a large marine embayment covered much of southern Utah, opening westward to the sea. On the coastal plain and in mudflats near the sea, successive layers of sand and mud accumulated, to become the dark red sandstone, siltstone, and shale of the Moenkopi Formation. Many of these rocks display well-preserved ripple marks, mud cracks, reptile and amphibian trackways, salt crystal impressions, gypsum (from evaporation of sea water), and alternation of mudstone, siltstone, and sandstone. Farther west a few limestone layers contain fossils of marine shellfish. The formation is nearly 1000 feet (300 meters) thick just west of the visitor center; it thins southeastward and finally disappears altogether. Sand, silt, and mud that make up the Moenkopi strata in this area came from highlands in western Colorado and to the south.

Above the Moenkopi Formation, the green and purple Chinle Formation was also partly deposited by water—probably fresh water. It contains a high proportion of volcanic ash decomposed into bentonite, a combination of clay minerals that swell when they are wet and shrink when they are dry. The puffy, crumbly surface produced by these changes is easily blown away by wind or washed away by rain.

The bentonite tells us that there must have been volcanoes somewhere in this area in Triassic time. And eruptions must have been voluminous; bentonite-rich parts of the Chinle Formation spread across southern Utah and northern Arizona, notably into Arizona's Painted Desert and Petrified Forest National Park. The Chinle Formation also includes several layers of hard, ledge-forming sandstone and conglomerate derived, it is thought, from highlands in central Arizona.

As volcanism died down in late Triassic time, a sandy desert developed here, one that rivals the modern Sahara and Arabian Deserts. Swept by winds, dunes built up across the land, later to become the Wingate Sandstone. Above the Wingate cliffs are thinner river-deposited sandstones separated by horizontal bands of fine siltstone together called the Kayenta Formation.

In Jurassic time, roughly 190 million years ago, dunes built another massive deposit, the Navajo Sandstone, which now rides the crest of Capitol Reef and shelters the many waterpockets of this park.

Rocks younger than the Navajo Sandstone appear in ridges and valleys along the east side of Waterpocket Fold. Limestone, sandstone, mudstone, and gypsum of the Carmel Formation suggest several incursions of the sea. The Entrada Sandstone, looking rather like the Moenkopi Formation on the west side of the fold, formed on floodplains and tidal flats; the harder gray Curtis Formation above it represents a brief incursion of the sea. Above them the Summerville Formation represents a tidal-flat environment again. The youngest Jurassic unit is the Morrison Formation, its bentonite-bearing shales as colorful as those of the Chinle Formation west of Capitol Reef. Fossil dinosaur bones have been found in the Morrison Formation here and elsewhere.

In Cretaceous time the land sank below sea level again. A thin layer of near-shore sandstone—the Dakota Sandstone—underlies a 3000-foot (1000-meter) sequence of gray marine Mancos Shale. The Dakota Sandstone in places contains rounded pebbles that indicate strong wave or current action; elsewhere it contains fossil oyster shells. East of the park boundary these rock layers flatten out perceptibly, and the gray Mancos Shale, weak and easily eroded, weathers into badlands. Still farther east, forming a mesa, are beach and near-shore deposits of the Mesaverde Group.

As the Mesozoic Era drew to a close, the Rocky Mountains were rising to the east, soon to be followed by the Wasatch Range to the west. These mountains brought many changes to this area.

Cenozoic Era. As the Colorado Plateau pushed upward, probably in fits and starts, its individual blocks, some raised higher than others, remained essentially horizontal. On the highest plateaus, erosion bit most deeply, stripping away many of the sedimentary strata. Hard limestone layers and well-cemented sandstone remained as plateau surfaces or, along folds and faults, stood as lines of cliffs.

There are no Tertiary sedimentary rocks in this park. Some undoubtedly once lay across this area: soft lake siltstone and limestone similar to that at Bryce Canyon National Park. But like the thick Cretaceous layers, they long ago eroded off Capitol Reef.

Black lava boulders on Johnson Mesa, near the park headquarters, tell of Tertiary volcanism and Pleistocene glaciation west and north of the park. The boulders, some of them 3 feet or more in diameter, are composed of 20-million-year-old basalt quarried about 25,000 years ago by Ice Age glaciers on high plateaus to the west. Streams gushing from melting glaciers tumbled the boulders downstream, until they were blocked by the partial barrier of Capitol Reef. Some were conveyed through the Fremont River's narrow canyon and dropped on the east side of the reef.

BEHIND THE SCENES

Capitol Dome. High on the crest of Capitol Reef just north of the Fremont River, Capitol Dome and several other dome-shaped summits are shaped in pale gray or yellowish Navajo Sandstone. Sweeping cross-bedding of the dune sandstone marks the domes and rocky uplands surrounding them. Narrow creases that run down the sides of the domes are channels eroded by rain.

Capitol Gorge. At the western, upstream end of this narrow gorge, Wingate Sandstone cliffs, tapestried with lichens, mineral stains, and glossy, blue-black desert varnish, rise from the streambed. The vertical walls also display large-scale, dune-style cross-bedding.

Pits and hollows of honeycomb weathering are common here. Starting with tiny depressions caused by solution of the calcium carbonate that holds the sandstone together, these pits are partly the work of wind, which helps to deepen them by whirling the loosened

Eons of erosion by wind, water, and frost have shaped the deep gorges and high, barren summits of Waterpocket Fold. Ledges and slopes are in the Kayenta Formation, with Navajo Sandstone above and Wingate Sandstone below. Ray Strauss photo.

sand grains and eventually removing them. Nearby, deep, smooth scallops in the rock wall are remains of potholes created by swirling stream water that spun rock against rock. The stream later sliced the holes in two as the canyon deepened.

Farther into the gorge, a Pioneer Register records early travelers who engraved their names in the desert varnish that coats the rock. (Some of these signatures are out of reach even from a wagon seat, and are thought to have been carved by pranksters lowered over the cliff by ropes.) Storm waters funneling through the narrow canyon frequently change the level of its sandy floor. Between 1880 and 1962 a road through Capitol Gorge was the only highway between Torrey and Hanksville; road repairs were necessary after every storm.

Still farther east, the Wingate Sandstone plunges beneath the surface, and the Kayenta Formation and Navajo Sandstone wall the canyon. Because these units are a little more easily eroded, they stand in less formidable cliffs. In them several waterpockets, known as the Tanks, usually contain standing water from the last storm. They lie along rock joints that were natural pathways for water sheeting off rock surfaces above. As channelways deepened, water collected behind cross-bedding ridges in the sandstone. Small pools became larger as acids secreted by tiny organisms dissolved cementing material that held the sand grains together.

A trail from Capitol Gorge climbs to the base of the Golden Throne, one of the massive Navajo Sandstone monuments that crown Capitol Reef.

Cassidy Arch. The trail to Cassidy Arch climbs steeply from Grand Wash through the Wingate Sandstone, then levels off among rounded erosional hummocks of Navajo Sandstone. Swirling cross-bedding marks barren uplands; overhangs and blackened hollows are incipient arches and waterpockets. Cassidy Arch itself, a graceful rainbow, is one of these features.

Cathedral Valley. Situated at the northern end of the national park, this valley is east of Waterpocket Fold. Here, in a maze of mesas and cuestas, 2000-foot (600-meter) cliffs overlook a broad, eroded surface dotted with cathedral-like buttes. The road into the valley passes through the Bentonite Hills in the Morrison Formation, part of Utah's Painted Desert, to reach overlooks above Cathedral Valley.

Shaped by wind and water, the great amphitheater is ornamented with unusual erosional forms carved in rocks younger than those of Capitol Reef—the Entrada Sandstone, of Jurassic and Cretaceous age. Dikes, sills, and volcanic necks jut above the surface. A gypsum sink results from solution of underground gypsum.

Chimney Rock. Shaped in red mudstone and siltstone of the Moenkopi Formation, this high tower stands out from a promontory of similar rock. Both are capped with hard, sandy, pebble-filled Shinarump Conglomerate. As you can see from slopes on the other side of the road, the Moenkopi is not normally a cliff-former. But

At Chimney Rock, unusually rapid erosion creates vertical walls of red Moenkopi Formation mudstone, normally a slope-former. The highest cap is resistant Shinarump Conglomerate. Ray Strauss photo.

here, where patches of Shinarump Conglomerate protect it and rapid erosion undermines it, it stands in almost vertical walls.

A fault separates these walls from the mesa to the north, where the Shinarump Conglomerate is at the base of the mesa cliff. Displacement along the fault is about 165 feet (50 meters). Petrified wood is common in the Chinle Formation here—not the prized agatized wood of Petrified Forest National Park in Arizona, but a duller variety impregnated with less colorful varieties of silica.

Ripple marks, mud cracks, and raindrop pits mark rock slabs of the Moenkopi Formation. The dark red color of these rocks is due to oxidation of iron minerals after the rock was deposited. White or pale pink veins are gypsum.

Cohab Canyon. The trail to Cohab Canyon climbs steeply through soft purple and green shale of the Chinle Formation, rich in clay minerals derived from volcanic ash. It then skirts towering cliffs of Wingate Sandstone, marked with black, yellow-green, and white mineral stains. Pronounced vertical fractures govern the shape of the cliffs. Some rock slabs have broken away in arcs, in shell-like (conchoidal) fracture patterns.

Cohab Canyon's narrow defile has eroded headward from east to west, and drains into the Fremont River. Along the trail watch for sedimentary and erosional details: ripple marks on flat rock faces, potholes, honeycomb weathering, and liesegang rings—decorative concentric brown arcs and bands caused by precipitation of iron oxide from groundwater.

Fremont River. Flowing from Fish Lake in the Fishlake Mountains northwest of Waterpocket Fold, the Fremont River transects the entire sequence of east-tilted Mesozoic rocks, from Moenkopi Formation redbeds near the visitor center to gray Mancos Shale east of the fold. Utah Highway 24 follows the river's route. Rock layers of the Wingate Sandstone, Kayenta Formation, and Navajo and Entrada Sandstones are tilted most steeply at the west side of the fold, and progressively more gently to the east. Farther east are the Bentonite Hills of the Morrison Formation—Utah's Painted Desert—and a sloping cuesta capped with tan Dakota Sandstone. Drab gray hills of Mancos Shale lead to mesas capped with Mesaverde Sandstone.

The waterfall near highway mile 86–87 was formed when highway construction blocked a gooseneck bend in the river. Black lava boulders, brought here by streams draining Pleistocene mountain glaciers, occur along the river for some distance east of the narrow part of the canyon.

Goosenecks of Sulphur Creek. Viewpoints at the Goosenecks look down on the oldest rocks in the park: buff-colored Cedar Mesa Sandstone, a Permian dune deposit, and above it yellowish or tan layers of the Kaibab Formation, marine limestone also of Permian age. Above these rocks is the dark red Moenkopi Formation, on which the viewpoints are sited.

The Goosenecks are entrenched meanders established when Sulphur Creek flowed with a gentle gradient across loose, poorly consolidated gravel deposited here in Pleistocene time. As the river deepened its canyon through Capitol Reef, it retained its old meanders.

Red sandstone and siltstone blocks lie scattered on the surface here, tilted in every direction as softer mudstone layers beneath are washed or blown away. Because two sets of joints, at right angles to each other, cut through these rocks, many of the blocks are almost rectangular. Some are ornamented with 200-million-year-old ripple marks; some show polygonal patterns of mud cracks, small craterlike raindrop pits, or cube-shaped casts of salt crystals. The lower surfaces of some blocks are marked with unusual angular ridges that may be sand-filled fin marks scraped in soft mud by denizens of Triassic seas.

Grand Wash. In its western part Grand Wash is walled with 800-foot (250-meter) cliffs of Wingate Sandstone. Farther east, where this formation is below the surface, the canyon cuts into Navajo Sandstone. Both formations are patterned with subtly colored mineral matter, dotted with lichens, and streaked with desert varnish. Vertical joints effectively maintain the steepness of canyon walls.

At the Grand Wash Narrows the canyon walls, only 20 feet (6 meters) apart at their base, are 500 feet (150 meters) high. The wash goes all the way through Capitol Reef, and makes a nice walk if arrangements can be made for a pickup on the other side.

Hickman Natural Bridge. This natural bridge was formed by gradual deepening of a waterpocket, which eventually joined up with a natural alcove in the cliff below. Weathering and further erosion enlarged the space below the bridge. The bridge is in the Kayenta Formation.

The trail to Hickman Bridge climbs through Wingate Sandstone, then leads across slickrock slopes. Many erosional forms that characterize Capitol Reef can be seen close at hand from this trail: rounded highlands marked with whorls of cross-bedding, barren slickrock slopes merging downward with ever-steepening cliffs, deeply etched overhangs shaped into small natural bridges. Capitol Dome and its neighboring summits rise nearby.

Notom Road. Leaving Utah Highway 24 just east of the national park, this route runs south along the east side of Waterpocket Fold. The road passes pinnacles of Entrada mudstone, then follows valleys eroded in

The oldest rocks in the park—the Permian Cutler Sandstone and Kaibab Limestone—are visible in canyon depths at the Goosenecks of Sulphur Creek. Ray Strauss photo.

tilted Jurassic and Cretaceous strata, curving in a gentle S parallel to the curve of Waterpocket Fold. To the west, Capitol Reef contains hidden waterpockets and natural bridges. To the east, a stockadelike ridge of Dakota Sandstone crowns colorful Bentonite Hills in the Morrison Formation. Resistant gravel and sand beds in the Morrison Formation contain petrified wood.

The Henry Mountains to the east, though outside the national park, are very much part of the geologic scene on this side of Capitol Reef. They consist of a cluster of laccoliths, igneous intrusions that pushed overlying sedimentary layers up into a dome. Except around the mountain base, the sedimentary rocks that once covered them are now eroded away. The ridge below the Henry Mountains exposes Cretaceous rocks: gray, slope-forming Mancos Shale and above it the Mesaverde Group.

South of Cedar Mesa the road rises to the Bitter Creek Divide, and south of there it closely parallels the Dakota Sandstone ridge, where fossil oyster shells are abundant. (No collecting within the national park, please.) Navajo Sandstone exposures to the east are marked with caves and alcoves, cross-bedding whorls, and many ravines and gorges. The dark red rock along their base is the Carmel Formation, above the Navajo Sandstone.

The Burr Trail branches from this route north of The Post. It offers first-class views of Waterpocket Fold and the valleys parallel to it, gives access to Muley Twist Canyon and high parts of the fold, and crosses the northern end of Grand Staircase–Escalante National Monument.

Scenic Drive. Winding through dark red-brown sandstone, siltstone, and mudstone of the Moenkopi Formation, this drive skirts the base of the Wingate cliffs and gives access to Grand Wash and Capitol Gorge. Along the road the Moenkopi Formation is spangled with shiny slabs of selenite, a form of gypsum, broken from veins that lace this rock. Elsewhere, red sandstone and siltstone are decorated with ripple marks and mud cracks.

East of the road, the Chinle Formation appears in purplish gray slopes ridged by thin, hard ledges of conglomerate. As it erodes it undermines the Wingate cliffs. Large slabs fallen from the 1000-foot (300-meter) wall may be marked with broad wind-formed ripple marks. Regulated by joint patterns, the cliff in places is decoratively fluted.

Farther south, the Moenkopi and Chinle Formations are separated by a resistant but discontinuous conglomerate layer, the Shinarump Conglomerate, considered part of the Chinle Formation. Erosion has carved many unusual shapes in this steeply tilted, pebble-filled rock layer.

OTHER READING

Collier, Michael, 1987. *The Geology of Capitol Reef National Park*. Capitol Reef Natural History Association, Lorraine Press.

Moore, Chris, 2001. *A Guide to the Natural Arches of Capitol Reef National Park*. Arch Hunter Books.

Olson, Virgil J., et al., 1990. *Capitol Reef: The Story Behind the Scenery*. KC Publications.

CHACO CULTURE NATIONAL HISTORICAL PARK

ESTABLISHED: 1907 as a national monument, 1980 as a national historical park with 33 outlying Chaco Culture Archaeological Protection Sites; in 1995, six more Protection Sites were added.
SIZE: 53 square miles (138 square kilometers), 22 square miles (53 square kilometers) of Protection Sites
ELEVATION: About 6000 feet (1830 meters) at visitor center
ADDRESS: PO Box 220, Nageezi, New Mexico 87037-0220

STAR FEATURES

☉ The ruins of prehistoric pueblos scattered in or near Chaco Canyon, where geology, archaeology, solar astronomy, and climatology come together to define a highly developed prehistoric culture.

☉ Cliffs of Cretaceous rock, well exposed and with weathering characteristics typical of a desert environment.

☉ A stream valley whose deposits reveal a long history of cyclic erosion and valley filling. Late stages of the erosion cycles had a pronounced impact on local inhabitants.

☉ Visitor center, museum, introductory movie, evening programs and guided tours, trails with guide leaflets describing the ruins.

SETTING THE STAGE

In and near flat-floored, cliff-walled Chaco Canyon are sites of numerous pueblos constructed between A.D. 900 and 1100 by prehistoric peoples whom archaeologists call the Anasazi. The canyon is walled with the Cliffhouse Sandstone, part of the Mesaverde Group—a Cretaceous rock deposited 70 to 80 million years ago near the shore of a shallow sea. The rock contains fossil shellfish, coal, shale, clay, and gypsum, all of which, with the sandstone used for building, played parts in the lives of the early inhabitants.

The large pueblos that occupy a 20-mile stretch of Chaco Canyon were apparently the center of a complex, largely agricultural culture involving at least fifty-nine pueblos, connected by 200 miles of roads. Fine stonework, black-and-white pottery made of local clays, and the use of coal, shale, and gypsum in jewelry and ornaments characterize the pueblo people. Chaco's inhabitants, moreover, established a unique "sun dagger," with a slender sliver of sunshine illuminating spirals carved in a rock cliff on the face of Fajada Butte—an accurate and still functional indicator of solstices and equinoxes.

Though the San Juan Basin, the wide desert in which the ruins occur, is lower than the plateaus to the west, it nevertheless is part of the great stable raft of the Colorado Plateau. The area drains northward, toward the San Juan River, but is given the name "basin" because sedimentary layers sag downward toward its center.

Chaco Canyon is floored with soft gray silt, visible in the walls of the present Chaco Arroyo, and with windblown sand, some of which covered the ruins of the Anasazi villages and sheltered them from centuries of erosion. Despite this protection, the ruins show the effects of natural processes: gradual disintegration of wooden roof beams and mud mortar used in building, partial destruction by earthquakes and rockfalls, the tumbling of a great block of sandstone near Pueblo Bonito (despite early attempts to stabilize it).

Chaco has recently been named by the World Monuments Fund as one of the hundred most-threatened monuments in the world. Deterioration of the site, in part due to natural processes such as erosion and earthquakes, is also due to rapid development of the San Juan Basin's petroleum resources, with road building, an

Pueblo Bonito was built on the floodplain of Chaco Wash, within easy reach of water. The wash now flows in the deep gully in the middle distance. Halka Chronic photo.

influx of people, and accelerated water use, which lowers the water table and increases erosion. Ruins that lie outside the national monument are particularly vulnerable to human influences.

GEOLOGIC HISTORY

Mesozoic Era. Pre-Cretaceous history of this region is probably similar to that of the rest of the Plateau country—repeated incursions of a western sea during the Paleozoic Era, and mostly nonmarine sedimentation in Triassic and Jurassic time. In Cretaceous time the sea again advanced across the land, coming this time from the east, depositing the thick sandstone and shale sequence of the Mesaverde Group. The youngest unit of the Mesaverde Group, the Cliffhouse Sandstone, forms the walls of Chaco Canyon. A near-shore sandstone deposited as the Mesaverde Sea retreated eastward, the unit represents the varied environment one would expect along a shore: In addition to an abundance of sand, it contains coal, shale, clay, and gypsum. Above it, toward the center of the San Juan Basin, are formations representing two subsequent advances and retreats of the Cretaceous sea. At the end of Cretaceous time the land rose once more, and ever since has been above sea level.

Cenozoic Era. Tertiary deposits to the east reflect uplift of the Rocky Mountains at the end of Cretaceous time. Downward warping of the San Juan Basin occurred at about the same time, with later regional uplift boosting the basin to its present 6000-foot (2000-meter) elevation.

Chaco Canyon was probably carved in Pleistocene

time, when glaciation to the north increased rainfall in this area, and downcutting by the San Juan River sped up erosion. Within the canyon the oldest Cenozoic deposits are late Pleistocene gravel. Studies of silty stream deposits above the gravel, and of fossil pollen that they contain, show that there have been several episodes of valley filling, separated by episodes of erosion, since Pleistocene time.

Pollen studies show an arid cycle from 5600 to 2400 B.C., when pine and pinyon forests that had earlier covered this area were greatly diminished. Pinyon woodlands began to reestablish themselves—

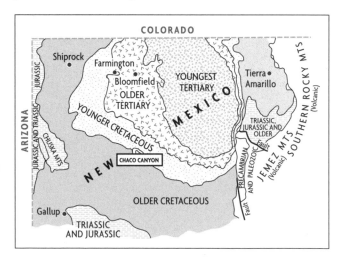

A geologic map of northwestern New Mexico shows the lopsided target of the San Juan Basin, with the youngest rocks as the bull's eye. Chaco Canyon is well off the center, in Cretaceous rocks that encircle the bull's eye.

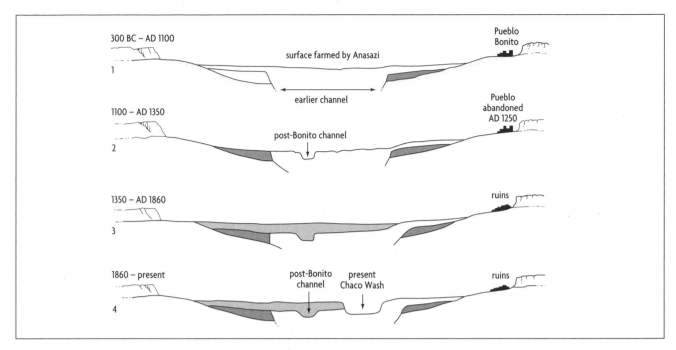

Chaco Canyon has seen several cycles of filling and trenching. Trenching of Chaco Wash may have caused abandonment of Pueblo Bonito and other Anasazi villages around A.D. 1250. Sections are not to scale.

reflecting increased rainfall—around 850 years ago.

When the great pueblos were occupied, between A.D. 900 and 1250, the climate was even drier and warmer than it is today. The Chaco Arroyo was probably a shallow streambed bordered by willows, sedges, and cottonwood trees. The pueblo inhabitants developed an elaborate water-control system that involved dams, canals, and reservoirs, with stone-lined irrigation ditches bringing water to farmed terraces. The system used both stream water from Chaco Creek and runoff from smaller creeks draining surrounding cliffs.

Why did the early Anasazi inhabitants, an obviously thriving people, abandon their pueblos? Poor land use may have played a part in the tragedy. And around A.D. 1200, subtle climate changes brought an increase in erosion. The shallow streambed deepened by as much as 10 feet (3 meters), possibly putting its waters out of reach of the irrigation system. Drought may also have been a factor. Recently, however, other hypotheses have been suggested—perhaps a religious change, increase in warfare, or other social upheaval influenced their departure. Within fifty years the pueblos were abandoned. Within another fifty years the channel began to refill with sand, the pinyon woodlands again expanded, and the climate

became similar to that of today—probably more livable than it was at the height of Anasazi occupancy.

Channeling of the present Chaco Arroyo dates from about 1860, and is part of a cycle of gullying recognized throughout the Southwest. Here at Chaco, erosion control initiated in 1935 to protect the ruins is responsible for accumulation of as much as 6 feet (2 meters) of new sediment in Chaco Arroyo.

OTHER READING

Anderson, Douglas, and Barbara Anderson, 1991. *Chaco Canyon: Center of a Culture.* Southwest Parks and Monuments Association.

Fisher, Reginald G., 1982. *Some Geographic Factors That Influenced the Ancient Populations of the Chaco Canyon, New Mexico.* Millefleurs Press.

Force, Eric R., et al., 2002. *The Relation of "Bonito" Paleo-Channels and Base-Level Variations to Anasazi Occupations in Chaco Canyon, New Mexico.* Arizona State Museum.

Vivian, R. Gwinn, and Bruce Hilpert, 2002. *The Chaco Handbook: An Encyclopedic Guide.* University of Utah Press.

COLORADO NATIONAL MONUMENT

ESTABLISHED: 1911
SIZE: 32 square miles (83 square kilometers)
ELEVATION: 4620 to 7101 feet (1408 to 2166 meters)
ADDRESS: Fruita, Colorado 81521-0001

STAR FEATURES

- ✪ Soaring cliffs and imprisoned canyons carved in rock layers of the Uncompahgre Plateau, a fault block lifted 6700 feet (2000 meters) relative to rocks below Grand Valley to the east.
- ✪ Precambrian metamorphic rock at least 1.7 billion years old, part of the faulted core of the uplift, and overlying Mesozoic strata that record a succession of seas, swamps, desert dunes, lush floodplains, and long periods of erosion.
- ✪ Visitor center, nature walks, self-guided trails, and evening programs, as well as roadside exhibits along scenic Rim Rock Drive.

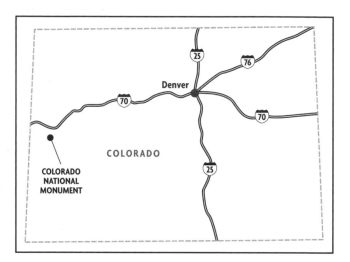

SETTING THE STAGE

This national monument lies at the northeast edge of the Uncompahgre Plateau, a fault block 125 miles long and 30 miles wide (200 by 50 kilometers), whose base is hard, dark gray Precambrian gneiss similar to that in the walls of the Black Canyon of the Gunnison and the Inner Gorge of the Grand Canyon. Above this ancient rock lie horizontally layered Mesozoic sedimentary rocks. Formations exposed here extend into adjacent

Jurassic and Triassic strata bend in a monocline across the fault that edges the Uncompahgre Plateau's east side. Halka Chronic photo.

states and appear in other national parks and monuments of the Colorado Plateau.

Landforms in Colorado National Monument are characteristic of regions where sedimentary strata are thick and horizontal and rainfall is relatively scarce. Because vegetation is sparse, erosion, largely by storm runoff and wind, has a free hand. And because resistant sandstone layers alternate with less resistant mudstone and siltstone, dizzying cliffs and sharply defined ledges alternate with fluted slopes and benches. Cliff height is dictated by the thickness and massiveness of sandstone layers, cliff verticality by their tendency to break away along vertical fractures.

With every storm, cliffs are further undermined and canyons bite deeper into the plateau. Through the centuries, parts of the plateau are little by little cut off, to become freestanding mesas. As erosion continues, mesas narrow into buttes and buttes into pinnacles, which eventually erode away completely. Streams, dry most of the year, show branching patterns that are governed to some extent by joints in the cliff-forming sandstone. Erosion along joints defines many features of the park: the Coke Ovens, Sentinel Spire, Balanced Rock, and others.

Many rock surfaces wear a thin black or dark brown coat of desert varnish. Others are marked with lichens.

The large fault along the northeast side of the Uncompahgre Plateau is typical of many Plateau country faults. While age-hardened Precambrian rocks that underlie the plateau were broken by the fault, sedimentary rocks above them stretched and draped over the edge of the fault like a thick bedspread, forming a prominent monocline. This downward bend of sedimentary strata along the flank of the plateau can be seen from both entrance roads and many places along Rim Rock Drive.

GEOLOGIC HISTORY

Precambrian Time. Dark reddish Precambrian gneiss and schist floor deep canyons of the monument and form the lowest juniper-dotted slopes. They are representative of the highly distorted rock that underlies most of the continent. Once sedimentary and volcanic deposits along an arc of volcanic islands, they are now altered almost beyond recognition. The only datable events in their history are their last recrystallization about 1.7 billion years ago, as the island arc welded to the continent under intense pressures and temperatures, and their intrusion by granite and pegmatite veins about 1.4 billion years ago.

The upper surface of the Precambrian rock appears as a straight horizontal boundary below the lowest red sediments. This surface represents an immensely long period of erosion, possibly as long as a billion years, during which the mountain ranges of which they are the roots were slowly stripped away. Final smoothing may have been accomplished by the sea, with the level surface forming just at wave base, where turbulence and wave action beveled underlying rocks to a remarkably even plane.

Paleozoic Era. As this ancient surface sank farther beneath the sea, layers of marine sedimentary rocks were deposited. Between 570 million and 300 million years ago, nearly 2000 feet (700 meters) of marine sandstone, shale, and limestone accumulated, much of it richly endowed with shells of marine organisms. We know the history of these rocks—the advances and retreats of the sea, the rise and fall of nearby land sources—only from adjacent areas, for no Paleozoic strata remain in the national monument.

In the middle of Pennsylvanian time, two large ranges rose in roughly the position of the present Rockies. The westernmost of these ranges geologists call Uncompahgria because it occupied about the same

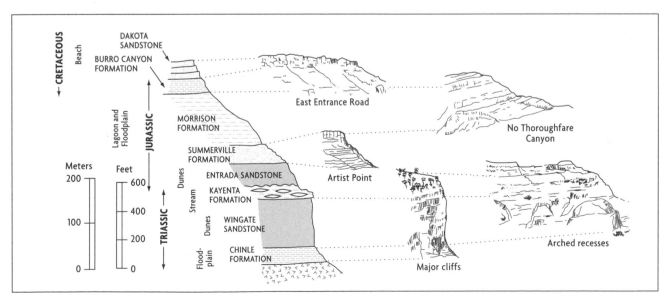

Stratigraphic diagram of Colorado National Monument

position as the present Uncompahgre Plateau. Exposed to the elements through the remainder of the Paleozoic Era and part of the Mesozoic Era, the marine sedimentary layers of Uncompahgria, much more easily eroded than the Precambrian rocks below, were washed off, right down to the old Precambrian surface.

Mesozoic Era. In the Uncompahgre region, Mesozoic deposition began soon after Paleozoic strata eroded away, so the Precambrian surface as we see it today probably corresponds fairly closely to the denuded surface formed at the end of Precambrian time. As the Mesozoic Era began, red sand and mud of a delta or floodplain spread across it. These sediments, now the Chinle Formation, appear near both monument entrance roads; watch for the red strata just above the Precambrian gneiss. The Chinle contains plant fossils, suggesting an environment supporting some vegetation.

Later, a creeping sea of windblown sand gradually covered an area extending southwestward and westward. The climate was arid, the scene definitely Saharan. The dune sands are now the Wingate Sandstone, more than 330 feet (100 meters) thick and the major cliff-former of the national monument. Visible on exposed surfaces are the long, diagonal laminae of the former

The Wingate Sandstone breaks away along vertical joints as at the right or, where joints are fewer, in soaring arches like that at the center. Foreground slopes and ledges are in dark "redbeds" of the Chinle Formation. Ray Strauss photo.

Cross-bedded Triassic sandstone of the Wingate Formation forms the cliffs of Colorado National Monument and such monoliths as Independence Monument. Ray Strauss photo.

dunes, some surfaces marked with wind-formed ripple marks, testifying to winds that swirled and eddied among the dunes 200 million years ago.

The Wingate cliffs are capped by the Kayenta Formation, a relatively thin layer of light-colored Jurassic sandstone and conglomerate, alternating with red and purple mudstone as streams began to weave among and across the desert sands. Well cemented with silica, the lowest sandstone of this unit acts as a hard caprock protecting the high Wingate Sandstone cliffs. The Kayenta Formation also forms the bench followed by parts of Rim Rock Drive. Where the caprock has been stripped away, unprotected Wingate Sandstone weathers into rounded domes, as at the Coke Ovens.

Above the Kayenta Formation is another gap in the record, expressive of continued sporadic uplift of Uncompahgria. Two Jurassic rock units known in adjacent regions, the Navajo and Carmel Sandstones, are missing here. Instead, clearly visible beside Rim Rock Drive north and south of the visitor center, is the record of another invasion of dune sand: the cross-bedded, salmon-colored Slick Rock Member, part of the Entrada Formation, which forms an easy-to-recognize, smooth-surfaced cliff. Horizontal layers within this rock are interdune deposits, silt and clay that accumulated in open spaces between individual dunes; some horizontal beds represent short-lived interdune ponds. At the time the Slick Rock dunes marched across the land, a shallow sea lay not far to the west, so these were, in reality, coastal dunes.

In the slope above the Slick Rock cliff are sandstone, siltstone, and mudstone of the Wanakah Formation, accumulated in shallow lakes and mudflats. Ripple marks on the flat flagstone layers suggest wave or current action. Thinner, smoother beds suggest deposition in quiet water.

Eventually, a new sequence of floodplain deposits accumulated, colorful gray, green, and red siltstones of the Morrison Formation. This unit points to a moister climate, development of shallow lakes and ponds, a rich supply of vegetation, and a correspondingly rich fauna. Many dinosaur skeletons have been found in the formation, including a monster that stood 60 feet (20 meters) high, discovered on the Uncompahgre Plateau south of the national monument. In addition to the skeletons, the Morrison Formation contains highly polished gastroliths, gizzard stones essential to dinosaur digestion. At the other end of the size scale, shells of freshwater snails and clams, as well as strange little water plants called charophytes, also occur in this formation.

Colorful siltstones of the Morrison Formation cover much of the plateau surface and appear as well along the entrance roads and at the base of the steep eastern scarp, where the upturned strata are carved into barren, bright-hued badlands.

Late in Jurassic time and early in Cretaceous time, rivers and streams deposited more sand and silt here, mixed with conglomerate derived from rising highlands to the southwest. These sediments, now the Burro Canyon Formation and the Dakota Sandstone, cap some of the higher parts of the national monument. Where they bend downward off the Uncompahgre Plateau they are visible along entrance roads near the monument boundary. The Dakota Sandstone, which forms a prominent hogback, collected along the shores of a Cretaceous sea as it retreated eastward. It is a thin but widespread layer of light tan sandstone that in places contains pebble zones and layers of black, coaly shale.

Rocks younger than the Dakota Sandstone, seen in gray slopes and palisades across Grand Valley, were eroded off the Uncompahgre Plateau as it rose to its present elevation. About 6500 feet (2000 meters) of Cretaceous marine shale and sandstone underlie Grand Valley and form the gray and yellow slopes of the Book Cliffs, the northern and northeastern boundary of Grand Valley.

Cenozoic Era. With the rise of the Rocky Mountains, sedimentation rates vastly increased. Above Book Cliffs are 8800 feet (2700 meters) of Tertiary deposits, products of floodplain, delta, and lake environments, that culminate in the Roan Cliffs. Both Cretaceous and Tertiary sediments once extended across a multistate region, but great quantities of these and other layers have since been eroded away, particularly from uplifted areas like the Uncompahgre Plateau. Later in Tertiary time, regional uplift in Colorado and adjacent states again revitalized erosion, with major rivers from the Rockies excavating broad valleys in soft Tertiary sediments.

OTHER READING

Baars, Donald L., 1998. *The Mind-Boggling Scenario of the Colorado National Monument: A Visitor's Introduction.* Canyon Publishers.

Scott, Robert B., et al., 2001. *Geologic Map of Colorado National Monument and Adjacent Areas, Mesa County, Colorado.* USGS Geologic Investigation Series I-2740.

DINOSAUR NATIONAL MONUMENT

ESTABLISHED: 1915, increased to present size in 1938
SIZE: 330 square miles (854 square kilometers)
ELEVATION: 4730 to 9005 feet (1442 to 2745 meters)
ADDRESS: 4545 E. Highway 40, Dinosaur, Colorado 81610-9724

STAR FEATURES

- ✪ A graveyard of the past, a rich dinosaur quarry with a protective exhibit building where visitors can see excavated bones. Portions of more than 300 dinosaurs have been found here.
- ✪ Scenic canyons of the Green and Yampa Rivers, unusual in that they cut through the Uinta Mountains rather than flow around them, the easier courses.
- ✪ Sedimentary, structural, and erosional features clearly exposed because of the arid climate.
- ✪ Dramatic Echo Park, where two rivers join amid a setting of steep-walled buttes and confined canyons.
- ✪ An interpretive program that includes the Dinosaur Quarry Visitor Center, the Monument Headquarters Visitor Center, naturalist talks and walks, slide shows, self-guided nature trails and drives, and trail and roadside exhibits.

SETTING THE STAGE

The canyon country where the Green and Yampa Rivers join is a rugged wonderland of pink and red sedimentary rock, an intricately eroded, almost inaccessible world little influenced by humans. Because average annual precipitation is only 9 inches (23 centimeters), vegetation is thin, and the rocks of this monument and the geologic structure of the Uinta Mountains are well displayed. Several impressive anticlines and synclines show up in the layered rocks. They, as well as a series of east-west faults, determine the shape of the Yampa Plateau and the Uinta Mountains. The oldest rocks are exposed in the walls of the Canyon of Lodore, bared by the Green River's downcutting. Successively younger rocks appear on the canyon rims and the flanks of the uplift.

What brought about the impressive scenery of this monument? And why do the Green and Yampa Rivers come together in the middle of a mountain range? There are a number of contributing factors:

- Many layers of sedimentary rock, some young and not very well consolidated, some far older and hardened into metasedimentary rocks.
- Development of the Uinta Mountains, an east-west range that consists of a single large anticline faulted along both edges and down its center as well.
- Almost complete burial of the eastern half of the Uinta Range early in Tertiary time, 65 million to 26 million years ago, by a thick blanket of silt, sand, and gravel.

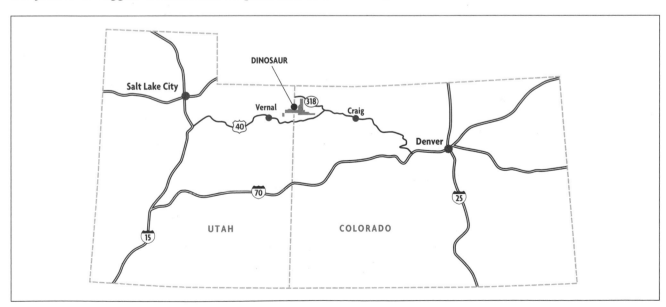

- Development of two sinuous rivers across the nearly level surface of this sedimentary mantle.
- Regional uplift of all of Colorado and large parts of adjoining states fairly late in Tertiary time.

Let us see how these factors fit into this area's history.

GEOLOGIC HISTORY

Precambrian Time. The story begins more than a billion years ago when sand, mud, and limestone layers were deposited in a sea-filled basin formed during the splitting apart of an ancestral continent. Time has hardened those sediments and colored them a deep red by gradually oxidizing iron-rich minerals within them. They are now the quartzite, slaty shale, and marble of the Uinta Mountain Group. They form the central core of the Uinta Mountains and appear in the walls of the Canyon of Lodore and in a few small, deep

Precambrian rocks in the Canyon of Lodore are still recognizable as sandstone, siltstone, and conglomerate. These old metasedimentary rocks extend from Canada to Arizona. Photo courtesy of USGS.

side canyons along the north edge of the Yampa Plateau. Though life existed when they were deposited, no fossils have ever been found in them here. In Glacier and Grand Canyon National Parks, corresponding rocks contain telltale remains of primitive algae and sponges, and imprints of jellyfish.

Paleozoic Era. Fossil-bearing sedimentary rocks of Paleozoic age tell us that this area was repeatedly flooded by shallow seas and that it rose above sea level for some time late in the era. Particularly applicable to the monument story are changes that took place in Pennsylvanian and very earliest Permian time, 320 million to 275 million years ago, when a pattern of rising and falling sea level left deposits of alternating sandstone and limestone. These gave way later to a pattern of coastal sand dunes edging a western sea. The sandstone-limestone sequence is now known as the Morgan Formation; the dunes have become the Weber Sandstone, the cross-bedded pink- and buff-colored rock whose barren slopes are so spectacular along the edge of the Yampa Plateau and in its many canyons. Its rounded, frosted, even-sized grains and large-scale cross-bedding assure us that it is a wind-blown, or eolian, sandstone.

Mesozoic Era. This era began with a new note—emergence of the land and an environment that alternated between lush vegetation on broad plains, dry deserts of drifting sand, and short incursions of the sea. During this era the region was low and much nearer to the equator than it is now. Red Triassic sandstone and siltstone, pebbly conglomerate, and limy mud were deposited on a low plain where lakes, ponds, and slow-flowing rivers were surrounded with vegetation. Some of the rocks contain petrified wood and fossil animal footprints.

During late Triassic and early Jurassic time, desert conditions predominated, possibly because new mountains along the coast of the continent were beginning to cut off moisture-bearing winds from the Pacific, as they do today. An incredible quantity of windblown sand came to rest here; this arid inland basin must have resembled parts of the present Arabian desert. Tan and pink, fine-grained, cross-bedded sandstone layers now stretch from western Colorado clear to Zion National Park in southern Utah. Though stream and beach deposits are often cross-bedded too, the particularly large cross-bedding in these sandstones and their rounded, pitted, uniform grains identify them as dune sands. Horizontally bedded sandstone and mudstone containing fossils of marine shellfish, evidence of brief incursions of the sea, separate several formations of dune sand.

Late in Jurassic time a rainbow-colored sequence of sandstone, conglomerate, shale, and freshwater limestone was deposited over an equally broad region, a wide floodplain at times studded with short-lived, shallow lakes. And in these sediments, now the

Morrison Formation, most of the dinosaur bones of this national monument were deposited and preserved.

Though dinosaurs in abundance have been found at only one site within this monument, in tilted layers on the south side of Split Mountain, they have been unearthed from the same formation at many other localities in Utah and Colorado. A sauropod skeleton recovered from Cretaceous sediments above the Morrison Formation is notable because it has a complete skull, rare among sauropods, and because it is Cretaceous rather than Jurassic in age. The giant reptiles, as well as their small relatives, may have perished during seasonal floods.

Later in Cretaceous time seas swept this area again, coming this time from the east. Layers of beach sand, lagoon shale and coal, and marine shale overlie the Morrison and its buried dinosaurs. Many fossil shellfish have been found in the thick marine shale: oysters, a variety of snails and clams, and beautiful ammonites, relatives of the octopus and the chambered nautilus. Cretaceous rocks now edge the Uinta Uplift and make narrow, upturned hogback ridges outlining the mountain front.

Cenozoic Era. Shortly after the end of the Mesozoic Era, mountain building that had started in the west with the Sierra Nevada and the Utah ranges crept eastward, forcing the sea to withdraw from the center of the continent. About 65 million years ago the Rocky Mountains (including the Uinta Mountains) began to push upward, isolating a spacious interior basin between the newly formed Rocky Mountains and other mountains to the west. Erosion of the Rockies contributed thousands of feet of sedimentary waste to this basin, enough cobbles, gravel, sand, and lake sediments thick with volcanic ash to cover all but the highest peaks of the Uinta Range. As time went on, two rivers established courses across the plain of Tertiary sediment. The ancestor of the Green River flowed east for a time into the Mississippi drainage but was diverted by continued rising of the Rockies and now flows south as a tributary of the Colorado River. The Yampa developed as a tributary of the Green. Both rivers found going pretty easy on the featureless blanket of Tertiary sediment, and they swung about lazily in looping meanders.

Strengthened by widespread uplift in Miocene and Pliocene time, the rivers finally began to carve down into the thick sediments, still following snakelike meanders inherited from their leisurely past. As they eroded more deeply, they eventually reached the buried mountains. Since by then they were completely imprisoned in valleys of their own making, they had to continue to carve on down into the hard rocks of the Uintas. In the last 12 million years they have lowered their beds at least 2000 feet (600 meters), retaining their former bends as incised meanders, keeping staunchly to their courses despite the hard Precambrian rock that cores the Uinta Uplift. Meanwhile, their tributaries cleaned away surrounding Tertiary sediments, leaving just a few remnants to remind us that these sands and gravels at one time covered most of the range.

Pleistocene climate changes that brought glaciation to North America and Europe scarcely touched this area, though valley glaciers developed in the Rockies, including the western part of the Uintas. Heavy Ice Age rains, coupled with melting from the glaciers, must have increased the volume of the rivers, speeding up the canyon-cutting process by contributing meltwater and broken rock and sand, an abrasive combination.

BEHIND THE SCENES

Canyon of Lodore. Named by geologist John Wesley Powell during his venturesome first trip down the unexplored Green and Colorado Rivers in 1869 (see sidebar), this canyon cuts deeply into the core of the Uintas. It begins at Browns Park, where the youngest rocks in the monument, the Browns Park Formation, lie next to the oldest, Precambrian quartzite of the Uinta Mountain Group. Hard red quartzite walls most of the canyon. Altogether there are about 20,000 feet (6000 meters) of this ancient rock which, with overlying younger strata, represent a time span of nearly a billion years.

Above the Precambrian rocks, visible from Harpers Corner or along the river downstream from Limestone Draw (whose name foretells a change in rock type), are Cambrian sandstone layers. They are separated from the Precambrian rocks by an unconformity, a break in sedimentary sequence between rocks deposited in Precambrian time and those deposited in Cambrian time. This break represents a time of erosion rather than deposition—many millions of years of uplift and bending and folding and beveling before Cambrian seas crept across the region. And above the Cambrian layers, above another but less profound unconformity, Mississippian limestone forms narrow bands of stepped gray cliffs. Where the Green and Yampa Rivers come together, they flow beneath cliffs of the wind-deposited sandstone of the Pennsylvanian Weber Formation.

Dinosaur Quarry. Here fossil dinosaur bones lie in sandstone layers in the Morrison Formation, a colorful shale and sandstone sequence deposited in Jurassic time, between 155 million to 147 million years ago, long before development of the Uinta Mountains. The rocks tell us of a broad floodplain with abundant savanna-like vegetation of conifer trees with a varied understory of cycads and ferns, and branching, usually sluggish streams. Plant-eating dinosaurs feasted on the rich vegetation, and meat-eaters preyed upon their fellows. As streams shifted, flooded, and shifted again, some of the dinosaurs perished and were buried in the shifting channel sand. Probably their bodies were washed and tangled and even broken apart by

THE PATHFINDER

When we return to camp at noon the sun shines in splendor on vermilion walls, shaded into green and gray where the rocks are lichened over; the river fills the channel from wall to wall, and the canyon opens, like a beautiful portal, to a region of glory. This evening, as I write, the sun is going down and the shadows are settling in the canyon. The vermilion gleams and roseate hues, blending with the green and gray tints, are slowly changing to somber brown above, and the black shadows are creeping over them below; and now it is a dark portal to a region of gloom—the gateway through which we are to enter on our voyage of exploration tomorrow. What shall we find?

John Wesley Powell wrote these words in 1869 just before entering, with nine companions in four large rowboats, what is now the northern part of Dinosaur National Monument. The names they assigned to geographic and geologic features are still used today: the Canyon of Lodore, Flaming Gorge, Disaster Falls (where one boat was wrecked), Echo Park, and many others. Their trip lasted three months and ended at the west end of Grand Canyon.

Despite his lack of a formal geologic education, Powell excelled at examining and describing geologic features along his route. He went on to found the U.S. Geological Survey, and served as its second director.

the stream currents before they came to rest in shallow bends and bars. The dinosaur quarry represents one such deposit—a concentration of bones in one sandbar. The sand that buried them is now sandstone, the bones are fossilized by gradual cell-by-cell addition of silica, and both sandstone and fossils are tipped up steeply at the edge of the Split Mountain anticline.

More than twenty complete dinosaur skeletons, as well as many incomplete ones, have come from this quarry since its discovery in 1909. They represent fourteen species of dinosaurs, including *Apatosaurus,*

Diplodocus, Allosaurus, Stegosaurus, and others whose names are less familiar. Fossil turtles, crocodiles, frogs, mammals, and freshwater clams occur here too.

Excavating fossil bones involves careful cleaning of their upper surfaces, painting them with penetrating varnish that strengthens and protects them, and coating them with thick layers of burlap dipped in plaster of paris. After the plaster hardens, rock beneath the bones is carefully removed, the plaster-and-bone block is turned over, and the same treatment is applied to the undersurface. Entire skeletons and hundreds of individual bones were completely quarried in this way and shipped to museums to be chiseled out of their plaster casts and studied or exhibited.

Bones are no longer being taken from this quarry. But some have been cleaned off as far as possible and soaked with preserving varnish, so that now they stand out in bold relief from the surrounding rock. In this way they give monument visitors insight into geologic methods as well as a view of a remote page of the geologic story.

Echo Park. Whether you see it from above, from viewpoints along the road to Harpers Corner, or from river level, Echo Park holds special fascination. Steamboat Rock's sharp, jutting prow dominates the confluence of the two rivers. The Yampa Fault, a long east-west fault that edges Blue Mountain and the Yampa Plateau, shows thousands of feet of displacement. Steamboat Rock and the other pink and peach-colored cliffs surrounding Echo Park reveal wide dune-style cross-bedding in the Weber Sandstone. The same formation occurs on the summit of Blue

Jumbled dinosaur bones are exposed on a steeply inclined surface of the Morrison Formation at the Dinosaur Quarry. Halka Chronic photo.

Mountain, 3000 feet (1000 meters) higher than the rivers near their confluence. A smaller fault block, dropped only half as far, creates the stairstep of Iron Springs Bench between Echo Park and the south canyon rim. The road to Echo Park takes advantage of the break in the cliffs along the Yampa Fault, and of the stairstep level provided by Iron Springs Bench, to reach the bottom of the canyon.

The Mitten Park Fault, which crosses the Green River just north of Steamboat Rock, is another thrust fault, bordering this area to the north. Its great up-swing, visible best from Harpers Corner, is one of the most dramatic in the monument.

Many rocky cliffs in this area are thinly coated with dark iron and manganese minerals drawn to the surface of the rock by thousands of years of desert sun and occasional moisture. Prehistoric Indians occasionally chipped through this desert varnish to the light-colored rock beneath, engraving long-lasting petroglyphs.

Harpers Corner. The road to Harpers Corner, in the Colorado portion of the monument, passes through a rock sequence that spans 250 million years of geologic time. Starting in Cretaceous rocks near Highway 40, it encounters successively lower, older strata of Cretaceous and Jurassic age. Beyond Mud Springs Draw

the road is bordered by cross-bedded pink and buff Weber Sandstone, laid down in drifting dunes of an ancient desert. Its sweeping slopes, almost devoid of soil and vegetation, are in places capped with patches of red conglomerate and sandstone and white volcanic ash of the Browns Park Formation, at only 20 million years old the youngest geologic unit in the monument.

Near where the road enters the main part of the monument, it crosses the Yampa Fault. Cliffs that edge the Yampa Plateau mark the fault's position. Rocks that are south of the fault on Blue Mountain are at water's edge near the Green and Yampa River confluence.

Viewpoints look out over the rocky wonderland of the Yampa and Green River Canyons, downfaulted between the Yampa Fault on the south and the Mitten Park Fault on the north and west. Twisting, incised meanders of the Yampa River can be seen, as can the lower end of the Canyon of Lodore along the Green River. Iron Springs Bench, the level area south of and above Echo Park, is a smaller fault block that dropped only about half as far as the rest of the downfaulted block.

Just beyond Iron Springs Bench Overlook, the road rises onto Harpers Plateau, climbing up across the sharp south edge, which marks the position of the Mitten Park Fault. Harpers Corner offers splendid

Incised meanders of the Yampa River date back to a time when the river wound across a wide lowland plain of Tertiary sand, silt, and gravel, most of it washed from the Rocky Mountain ranges to the east. W. B. Cashion photo, courtesy of USGS.

Ridges in the foreground echo the arch of Pennsylvanian Weber Sandstone across Split Mountain Anticline. Lucy Chronic photo.

views east up the canyon of the Yampa River, north to the Canyon of Lodore, and west to Whirlpool Canyon, Island Park, and Split Mountain. It is surfaced with small patches of Tertiary gravel, cobbles, and sand deposited before canyon-cutting began.

Island and Rainbow Parks. Accessible only by river or rough dirt road, this open region is northwest of Island Park Fault. Rocks that floor it are much younger than those in the canyon slopes across the river, on the upthrown side of the fault. Colored shales that give Rainbow Park its name are parts of the Morrison Formation, the same rock unit that contains the Dinosaur Quarry. Free here of confining canyon walls, the Green River braids its way around a large bend that is probably similar to meanders originally formed on the blanket of Tertiary sediment.

Split Mountain and Split Mountain Canyon. As one of the anticlines fringing the south side of the Uinta Range, Split Mountain displays arched layers of sedimentary rock that range in age from Mississippian in the depths of Split Mountain Canyon to Cretaceous in hogbacks that border the mountain itself. Split Mountain is oval-shaped, long in the east-west direction, and "split" diagonally by the Green River as it adheres to a course established in Miocene time. Its tributaries, revitalized by uplift and swollen by Ice Age storms, cleared away loose Tertiary sand and gravel that had completely concealed the surrounding highlands. The south side of Split Mountain and the mouth of its canyon can be seen from the campground; river-runners see both sides of the anticline.

Yampa Canyon. Good views of the Yampa region can be had from the Round Top fire lookout on Blue Mountain and from viewpoints on the road to Harpers Corner. Inner parts of the canyon are accessible by river or, in dry weather, by rough dirt roads.

The Yampa is bordered on its course through the national monument by Pennsylvanian rocks belonging to the Morgan and Weber Formations. The Morgan is a series of flat-bedded sandstone and limestone layers, some of them red. It is a marine deposit and contains fossil shellfish. The Weber Sandstone walls much of the river's narrow inner canyon, in places overhanging the water as pink cliffs marked with swaths of desert varnish. The river follows a twisting, canyon-confined course through meanders inherited from Tertiary time.

OTHER READING
Hagood, Allen, and Linda West, 1990. *Dinosaur: The Story Behind the Scenery.* KC Publications.

Hansen, Wallace, 1997. *Dinosaur's Restless Rivers and Craggy Canyon Walls: A River Runner's Guide to the Geology of a Classic Canyon Wilderness.* Dinosaur Nature Association.

Powell, John Wesley, 1876. *Report on the Geology of the Eastern Portion of the Uinta Mountains and a Region of Country Adjacent Thereto.* U.S. Geologic and Geographical Surveys of the Territories.

West, Linda, and Dan Chure, 2001. *Dinosaur: The Dinosaur National Monument Dinosaur Quarry.* Dinosaur Nature Association.

EL MORRO NATIONAL MONUMENT

ESTABLISHED: 1906
SIZE: 2 square miles (5 square kilometers)
ELEVATION: 7218 feet (2380 meters) at visitor center
ADDRESS: Rt. 2, Box 43, Ramah, New Mexico 87321

STAR FEATURES

✪ A cliff of Jurassic sandstone formed from ancient dunes, inscribed with prehistoric and historic petroglyphs and inscriptions.

✪ An unusual but dependable waterhole that for centuries has furnished plentiful water in an otherwise arid land.

✪ Erosional landforms that include Inscription Rock and nearby cliffs, with clefts and alcoves shaped by water, wind, frost, and gravity.

✪ Visitor center displays, evening programs, and a descriptive trail leaflet for walks around and over Inscription Rock.

SETTING THE STAGE

The Zuni Mountains of western New Mexico, an oval, dome-shaped uplift, have a central core of Precambrian granite, gneiss, and schist. Tilted layers of sedimentary rock that once blanketed the whole area now appear in concentric bands around the denuded core, the resistant ones shaping hogbacks and cuestas whose cliffs stand like rings of bleachers, the weak ones eroded into racetrack valleys between. This pattern, though easy to see from the air or on a geologic map, is obscured by its very size when seen from the ground, especially since parts of the sedimentary bands are concealed by lava flows. Prominent among the outermost rings is the Zuni Sandstone, curving around the southeastern end of the uplift and forming the cliffs of El Morro National Monument.

This sandstone, fairly soft and easily carved, capped by protective ledges of harder Dakota Sandstone, rises above a valley of weak, colorful shales of the Chinle Formation. Its long, sweeping cross-bedding and fine, evenly rounded, frosted quartz grains show that the Zuni Sandstone is a dune deposit. The darker, browner Dakota Sandstone above it is a mixture of pebbles and sand probably deposited by streams flowing across a coastal plain near the sea.

Inscription Rock and the bluffs near it (*el morro* means "the headland" or "the bluff") reveal many fea-

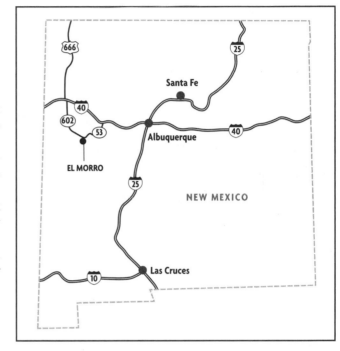

tures of both of these sandstones, and show weathering characteristics typical of semiarid lands. Along the trail at the base of the cliff, and on the trail over the top of the mesa, you will see examples of these processes. Watch for these features:

• Desert varnish—manganese and iron minerals cemented by microorganisms from dust (see sidebar on page 66).

• Arched alcoves that develop as blocks of unsupported rock break away and tumble from the cliff face.

• Deep horizontal clefts caused as descending groundwater flows horizontally along relatively impermeable layers of fine clay that mark the position of old interdune deposits.

• Vertical faces that result from weakening, breakage, and erosion along prominent vertical joints. In places the broken faces show arches of curving conchoidal (shell-like) fractures. Elsewhere they display exfoliation, the peeling off of thin scabs of rock in a continuous weathering process.

• Details of the uneven contact between the Zuni Sandstone and overlying Dakota Sandstone, a contact that represents a 30-million-year break in sedimentation. Channels along the contact are

commonly filled with wave-washed blocks of white Zuni Sandstone imbedded in the browner shoreline sands of the Dakota Formation.
- On the trail over the mesa, views into Box Canyon—cliff-walled and remarkably deep for a canyon with such a small drainage area. Joints in the Zuni Sandstone, as well as the varying hardness of the sandstone, seem to have provided an avenue for unusually rapid erosion.

The waterhole at the base of Inscription Rock was no doubt the main attraction here in prehistoric days as well as in the days of Spanish and Anglo-American exploration and settlement. It is not a spring, but a deep pool filled by rain and snowmelt draining from the mesa top down deep joint clefts in the sandstone. The names on the rock above the pool were carved at a time when a sandy beach extended along one side. After a rockfall filled the waterhole in 1942 (leaving a light-colored, relatively unweathered rock face to show where it broke away), both rock and sandbar were removed. The small dam is a modern addition, slightly increasing the size of the pool.

GEOLOGIC HISTORY

The geologic story here is no less interesting than the inscriptions on Inscription Rock, but very little of it can be deduced from El Morro alone. We must look to

The plunge pool at the base of El Morro's cliffs collects runoff from the mesa surface. Streaks of black lichens mark seepage channels in the rocks above. Halka Chronic photo.

ARCHES NATIONAL PARK

▲ *Eolian sandstone of the Navajo Formation forms the mesa surface near Elephant Butte and the Windows Section, where arches are carved in Entrada Sandstone. The distant La Sal Mountains rose as hot magma pushed upward in clustered laccoliths. Halka Chronic photo.*

▼ *Rock layers fracture as they bend over the anticline, with parallel joints that widen upward. Lucy Chronic photo.*

▲ *On the standing sentinels of Courthouse Towers, the Entrada Formation breaks away in great straight slabs, leaving abrupt cliffs rather than rounded arches. Subtle differences within the same formation lead to different styles of breakage. Lucy Chronic photo.*

BRYCE CANYON NATIONAL PARK

▲ *Differential erosion of slightly harder and slightly softer sediments results in strange, lifelike spires. Lucy Chronic photo.*

▲ *Hikers view aspects of Bryce not seen from the rim. Note the slanting scour marks of an earlier soil level at the base of the wall at the left. Halka Chronic photo.*

▶ *The dramatic escarpment at Bryce results from headward erosion of numerous streams draining a fault scarp along the east edge of Utah's Paunsaugunt Plateau. Lucy Chronic photo.*

CANYONLANDS NATIONAL PARK

▲ Where the Green and Colorado Rivers join, their separate waters flow briefly side by side. In this photo the Colorado is brown and the Green is green. Color depends on upstream storms that wash mud into one or both rivers. Lucy Chronic photo.

▼ A wide flat-floored vale near its southern end, the Devils Lane graben narrows northward to less than 9 feet (3 meters). Halka Chronic photo.

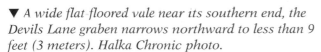

▲ A long, narrow graben, dropped neatly between nearly vertical faults, opens up like an inviting trail. Erosion widens closely spaced joints in the colorful Cutler Formation to create the mounds on either side. Lucy Chronic photo.

▲ Soft layers of siltstone beneath harder sandstone erode to form a balanced rock. Lucy Chronic photo.

CAPITOL REEF NATIONAL PARK

▲ Clays derived from volcanic ash expand when wet, hindering plant growth on the Chinle Formation; above, Wingate Sandstone forms spires of a "crown." Lucy Chronic photo.

▲ The Navajo Sandstone erodes into the white domes that give Capitol Reef its name. The term "reef" was used by early settlers to denote a barrier. Lucy Chronic photo.

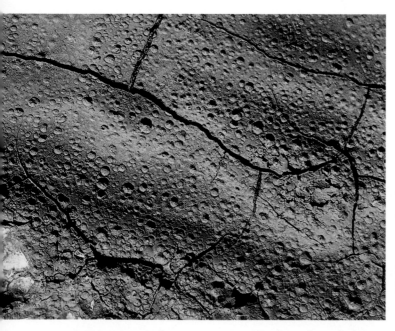

▲ Triassic raindrop impressions look as though they formed only yesterday. Halka Chronic photo.

▲ Volcanic boulders in this gully have washed from lava flows in nearby mountains. Lucy Chronic photo.

GRAND CANYON NATIONAL PARK

▲ Seen from Desert View, bright red Precambrian sedimentary rocks are beveled and covered by horizontal Cambrian strata. Halka Chronic photo.

◄ Wide, placid waters funnel toward the narrow Inner Gorge, where hard Precambrian rocks confine the river and increase its speed and turbulence. Halka Chronic photo.

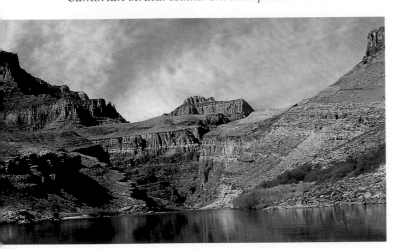

▼ Black lava cascades down canyon walls at Toroweap. The wide, flat bench of the Esplanade formed as Hermit Shale eroded back from the rim, causing retreat of cliffs of Coconino Sandstone and Toroweap and Kaibab Limestone. Lucy Chronic photo.

▲ Once the playthings of the untamed river, boulders such as these are no longer swept downstream. Jim Greslin photo.

GRAND STAIRCASE–ESCALANTE NATIONAL MONUMENT AND VERMILION CLIFFS NATIONAL MONUMENT

▲ *Tributaries of the Escalante River carve narrow slot canyons in Navajo Sandstone. During thunderstorms, debris-filled water rushes through the canyons, scouring out everything in its way and leaving the walls smoothed and rounded into sculptural forms. Lucy Chronic photo.*

▲ *Grosvenor Arch, eroded in Jurassic and Cretaceous sandstone, has a smaller window next to it. Lucy Chronic photo.*

▼ *Vermilion Cliffs rise above the stark plain of the Marble Platform. On hills at the base of the cliffs, strata tilt back toward the cliffs, showing that they slid downward along curving slide planes. Lucy Chronic photo.*

PETRIFIED FOREST NATIONAL PARK

▲ *Sections of petrified trunks, broken along joints, reveal their cellular structure and the colorful agate and jasper with which they are impregnated. Halka Chronic photo.*

▶ *Blue Mesa's brown slopes are paved with pebbles fallen from the conglomerate at the right. Many of the pebbles came from Triassic mountains in central Arizona. Blue, gray, and white banding marks volcanic ash layers. Halka Chronic photo.*

▲ *Cloud shadows accent the Painted Desert's slopes, colored with iron and manganese oxides. White deposits along dry watercourses are lime (calcium carbonate) and gypsum. Halka Chronic photo.*

ZION NATIONAL PARK

▲ *Zion Canyon's 2000-foot (610-meter) walls expose the great thickness of Navajo Sandstone. Lucy Chronic photo.*

▲ *In dry weather, Zion's sheer-walled Narrows invite hikers. They are out of bounds, though, during summer thunderstorm season. Felicie Williams photo.*

▲ *Weeping Rock's tears issue from seeps along the contact between porous Navajo Sandstone and impervious mudstone and siltstone of the underlying Kayenta Formation. Lucy Chronic photo.*

▲ *An opportunistic plant finds a place in a joint in the Navajo Sandstone. In a small way, the plant will participate in the ultimate wearing down of the rock. Lucy Chronic photo.*

the rest of the Zuni Mountains, and indeed to the entire Plateau region, for the rest.

Precambrian Time and Paleozoic Era. The granite, gneiss, and schist exposed in the Zuni Mountains tell of tremendous temperatures and pressures and alteration of preexisting rocks during Precambrian time, when this area collided with the tectonic plate that is now North America. After the long erosion at the end of Precambrian time, the area, then at the western edge of the continent, was covered by a succession of shallow Paleozoic seas. Flat-lying marine limestone, sandstone, and shale layers deposited in these seas are similar to those at Grand Canyon National Park, but here, periodic uplift caused many layers to be stripped away almost as soon as they were deposited. Only Pennsylvanian and Permian strata were preserved, lying directly on Precambrian rocks. These strata correlate with the uppermost layers at Grand Canyon.

Mesozoic Era. As the Mesozoic Era opened, the area lifted and the sea retreated. In lagoons, marshes, and rivers, and on gently shelving shores of Triassic time, thick layers of silt and mud were deposited. In Jurassic time these strata were covered with dunes of windblown sand that eventually became the Zuni Sandstone. Then the Cretaceous inland sea, coming from the east, swept the region, and in retreating left the brownish near-shore sand and mud of the Dakota Formation. Another advance left layers of dark gray marine shale throughout the Colorado Plateau and Rocky Mountain region. Near the end of Cretaceous time, about 70 million years ago, as the Rockies began to rise, the sea backed away, once more leaving near-shore sands with interlayered beds of plant material that would later become coal.

Cenozoic Era. In New Mexico the early part of the Cenozoic Era was a time of uplift, volcanism, erosion, and more volcanism—a time of continental extension or pulling apart, when deep faults that reached down into the molten material of the Earth's mantle brought volcanic activity throughout the Southwest. As the Zuni Mountains pushed up, erosion carved deeply into the sedimentary rocks that covered them. Volcanoes showered ash upon the land, and molten rivers of lava crept down stream valleys and across the eroded surface. Eventually the entire region, with all its valleys and mountains, was lifted to its present elevation. Uplift aided and abetted erosive forces and brought about renewal of volcanism. Rain and wind, helped by frost and melting snow of the Pleistocene Epoch's cooler, wetter climate, deepened canyons and valleys, sharpened cliffs and ridges. Desert winds and sporadic storm-fed torrents continue these processes today.

Within the last few seconds of geologic time, prehistoric peoples, Spanish conquistadores, and Anglo-

Names and dates are carved in soft Zuni Sandstone, formed in Jurassic dunes. Halka Chronic photos.

American explorers, soldiers, and settlers carved pictures or their names and the dates of their passing on the tablet of cliffs around the waterhole.

OTHER READING

Foster, Roy W., 1971. *Southern Zuni Mountains: Zuni-Cibola Trail.* Scenic Trips to the Geologic Past, no. 4. New Mexico Bureau of Mines and Mineral Resources.

Murphy, D., 1989. *El Morro National Monument.* Southwest Parks and Monuments Association.

Noble, David Grant, and Richard B. Woodbury, eds., 1993. *Zuni and El Morro Past and Present.* Ancient City Press.

GRAND CANYON NATIONAL PARK

ESTABLISHED: 1908 as a national monument, 1919 as a national park
SIZE: 1902 square miles (4926 square kilometers)
ELEVATION: 1800 to 9161 feet (613 to 2792 meters)
ADDRESS: PO Box 129, Grand Canyon, Arizona 86023

STAR FEATURES

- ✪ Geology's masterpiece, an awesome abyss unsurpassed for scenic grandeur. Grand Canyon's walls, with their orderly array of sedimentary layers lying horizontally above a beveled surface of much older rocks, unfold the pages of 2 billion years of Earth history.
- ✪ Erosional landforms geologically young and typical of regions of horizontal strata and desert climate.
- ✪ Two hundred and eighty miles (450 kilometers) of the Colorado River, the master sculptor, itself ruled by tectonic uplift, downdropping of faults along the edge of the Colorado Plateau, and the opening of the Gulf of California.
- ✪ Volcanic features that add dark overtones in western Grand Canyon and offer important clues to the canyon's history.
- ✪ Many trails, some with trailside signs interpreting the geology, roadside displays, nature walks, interpretive talks, shuttle bus tours (summer only) to viewpoints along the rim, visitor centers, and museums with displays of art, history, prehistory, and geology.

See color pages for additional photographs.

SETTING THE STAGE

One of the marvels of Grand Canyon is that it is here at all. Why should a river flow through a 9000-foot (3000-meter) uplift? For the Colorado River bisects the very heart of the Kaibab Uplift, and bisects it so thoroughly that the river flows in desert summer while pine and aspen clothe the cool plateaus above, so thoroughly that animals south of the canyon have evolved differently than those north of the canyon, so thoroughly that it is an effective barrier even to humans, who must drive 235 miles (380 kilometers) to see the canyon from the other side.

The great chasm trends roughly east-west. Because the surface of the uplift and the rock layers within it slope gently southward, regional drainage on the North

Rim flows toward the canyon, while that on the South Rim flows away from the canyon. The largest tributaries therefore have established themselves north of the river—Bright Angel, Shinumo, Kanab, and other creeks—and the Colorado River is much closer to the South Rim than to the North Rim. The river is 4460 vertical feet (1360 meters) and a little more than 2 horizontal miles (3.5 kilometers) from Grand Canyon Village on the South Rim, and 5940 feet (1810 meters) and 7.75 horizontal miles (12.5 kilometers) from Bright Angel Point on the North Rim.

Through the ages the Colorado River has cut downward through one rock layer after another, in a peculiar history that involves drainage patterns in Arizona, Utah, and Colorado. But its once turbid waters are tamed now by dams that retain floods and withhold the mud, sand, and gravel that used to come from far upstream.

Through much of the canyon's length the river is imprisoned by resistant rock of the narrow Inner Gorge. It does not swing widely in a broad valley, as do most major rivers. Canyon widening we must credit to other agents: storm-fed tributary streams, temperature changes, wind, and gravity. Summer storms, severe but

From the dark Precambrian gneiss of the Inner Gorge (lower left) to the white Permian limestone of the canyon rim, Grand Canyon offers an unparalleled view of much of the Earth's history. Halka Chronic photo.

brief, bring every ravine, every gulch to life, and churning rivulets plunge from cliff to cliff until the canyon is a world of waterfalls, many of them red with silt and mud from redrock slopes. The short-lived torrents carry silt, sand, and pebbles into tributary canyons. Reaching at last the lowest cliffs, they cascade into the Colorado and contribute their colorful loads to the river. The largest storms fill tributary canyons with tumbling torrents that tear loose and carry rock fragments of all sizes downstream toward the river.

Daily temperature changes in the canyon are sufficient to crack the rocks of its walls. On frosty nights, frequent along canyon rims, less common in its depths, frost forming in cracks and crevices wedges rocks apart. Angular slabs, some tiny, some tremendous, are by degrees pried away, and finally, perhaps only after many centuries of such frost action, they tumble and slide to the slopes below. Rockfalls and landslides widen the canyon under the influence of gravity. Those who have lived on the rim have heard their thunder; piles of angular boulders or patches of newly exposed, unweathered rock tell their story. Many rockfalls are caused by oversteepening of slopes and undermining of cliffs by wind and rain. Large backward-tilted blocks, common features in parts of the canyon, slid downward along curving slide surfaces. Some individual buttes in Grand Canyon now stand out from the rim because of fault motion down and away from the rim.

The colorful layers of sedimentary rock show different degrees of resistance to wearing-down and widening processes, making of the canyon walls magnificent illustrations of differential erosion. Well-cemented sandstone and limestone tend to break away into vertical cliffs and ledges, their heights corresponding to the thickness of the rock layers. Siltstone and mudstone erode into gentler slopes and benches. In eastern Grand Canyon one bench in particular

Dark, resistant metamorphic rocks—the Vishnu Metamorphic Complex—wall Grand Canyon's somber Inner Gorge. Ray Strauss photo.

At the east end of Grand Canyon, the Little Colorado River builds a rocky fan. A repeat performance, the delta will be swept away when more water is released through Glen Canyon Dam. Tad Nichols photo.

stands out: the wide Tonto Platform, which spreads a gray-green apron above the Inner Gorge. (Tiny trails that cross the Tonto Platform provide a measure of its breadth: They are 6 feet wide or more.)

Since the sedimentary strata lie in horizontal, alternately hard and soft layers, cliffs and slopes repeat themselves below every point and across every ravine and canyon, reaching symmetrically toward the dark, hard cliffs of unstratified igneous and metamorphic rocks of the Inner Gorge. Only in western Grand Canyon is there a major change of pattern. There the Tonto Platform disappears, and soft red siltstone and mudstone layers above the Red Wall erode back into a wide shelf known as the Esplanade.

Despite the symmetry, there are many minor changes from place to place in different parts of the canyon walls, as there are in rocks everywhere. Geologists find these changes particularly exciting here because rock layers can be traced visually for long distances, or followed by walking along well-defined cliff tops or individual ledges. Here, many ideas about geologic processes can be observed firsthand. Sandstone interlayered with shale or limestone visibly demonstrates fluctuations in sea level. River-deposited red siltstone giving way westward to shore deposits and eventually to marine limestone demonstrates differences in environment across coastal regions. Ancient channels and even more ancient islands are expressed in surrounding rock layers. Pebble-filled conglomerate on old erosion surfaces indicates the end of erosion and the beginning of deposition. And from bottom to top in the sedimentary layers, fossils reflect the evolution of life from primitive Precambrian algae through shell-bearing denizens of Paleozoic seas, from early fishes to lizardlike reptiles that left their footprints on Permian dunes.

Grand Canyon also furnishes unequaled examples of a primary geologic "law": In undisturbed sedimentary rock layers, the oldest rocks are at the bottom, and those above are sequentially younger and younger. An understanding of this precept adds to our understanding of geologic time and drops each layer of the canyon walls into place: the oldest near the river, the youngest at the rim.

Grand Canyon properly begins where the Little Colorado joins the Colorado, east of Desert View. It ends at Grand Wash Cliffs, the southwest boundary of the Colorado Plateau. The national park now includes Marble Canyon, upstream from the Little Colorado, and extends west to the stilled waters of Lake Mead. Between the Little Colorado and the base of Grand Wash Cliffs, plateaus rise and descend stairstep fashion, with faults and monoclinal folds controlling their shapes and height.

GEOLOGIC HISTORY

Precambrian Time. In Grand Canyon, Precambrian rocks fall into three rough age groups: the highly metamorphosed Vishnu Metamorphic Complex, which had its beginnings around 2 billion years ago; granite that intruded at various times; and a much younger group of sedimentary and volcanic rocks.

The Vishnu metamorphic rocks accumulated as sediments and volcanic rocks on the seafloor between volcanic islands that lay off the coast of the ancient North American continent 1.8 to 2 billion years ago. Studies of the sequence of rocks show that they were involved in at least two pulses of mountain building, and subjected to intense pressures and extreme temperatures at depths as great as 12 miles (20 kilometers) below the surface, thanks to major collisions between the continent and offshore islands or microcontinents.

Intrusion of the granite occurred in three phases, two of them during the Vishnu metamorphism. The third, around 1.5 billion years ago, was accompanied by large-scale faulting, particularly along north-south faults that may represent a period of rifting, or partial breakup, of the continent.

Mountains formed during metamorphism of the Vishnu metamorphic rocks and intrusion of the granite were leveled by erosion long before the younger Precambrian sedimentary rocks now visible in the canyon were deposited. These sedimentary rocks, along with some Precambrian volcanic rocks, appear in faulted, tilted wedges in the eastern part of Grand Canyon below and across from Desert View and in Bright Angel Canyon, as well as in several remote areas to the west. Sandstone, shale, limestone, conglomerate, and lava flows are still easy to recognize, even though they were deposited around a billion years ago. Some are imprinted with ripple marks and cross-bedding; others display relic mud cracks. A few of them contain some early traces of life (see sidebar).

After these rocks were deposited, probably about 800 million years ago, they were broken and tilted along earlier faults to form a series of fault-block mountain ranges. Then, during 300 million years of erosion at the end of Precambrian time, they and the older Precambrian rocks were trimmed away to an essentially horizontal surface, a peneplain now expressed by the Great Unconformity between Precambrian and Paleozoic rocks—an unconformity well defined not only in Grand Canyon but in many other parts of the world.

Paleozoic Era. In Grand Canyon the Great Unconformity is clearly exposed for hundreds of miles, and can be studied in greater detail than anywhere else on Earth. Hills and ridges of resistant Precambrian quartzite, islands in the Cambrian sea, in places protrude upward into Cambrian sedimentary rocks. Imbedded in Cambrian strata are blocks of broken rock torn from the islands by surging waves. Sea cliffs undercut by waves, even the flattened pebbles of shingle beaches, can be found along the Precambrian-Cambrian unconformity.

Cambrian strata here illustrate the detail to which geologic history can be worked out when rock layers are well exposed over long distances. Research shows that the Cambrian sea advanced in several pulses, with predictable changes in environment from the shore outward: conglomerate and coarse sandstone near the shore, finer sandstone and mudstone farther out, and finally limestone where the sea was fairly open. Small animals—trilobites, brachiopods, simple cone-shaped gastropods—burrowed in or crawled about on the sea bottom, leaving both trails and shells. Sponges, bryozoa, and various forms of seaweed flexed with tidal currents.

Toward the end of the Cambrian Period the sea retreated. No record of Ordovician or Silurian time exists here. Either no deposits formed, or they all later washed away. Shallow canyons eroded in the top of the Cambrian sequence are filled with thin layers of Devonian dolomite—purplish, gnarled, and sugar-textured—replaced westward by continuous bands of white and gray dolomite.

After yet another erosional period when the sea retreated and readvanced, the Redwall Limestone was deposited here in Mississippian time in two pulses of the continental sea. Thick, strong, and extremely widespread, the limestone contains fossil fishes and many small marine invertebrates such as brachiopods, corals, mollusks, crinoids (sea lilies), and trilobites. The Redwall sea was warm and moderately shallow—a perfect environment for the variety of plant and animal life evident from the fossil record. The rock is not really red at all, but bluish gray; in Grand Canyon its surface is stained with red iron oxide mud from the colorful shales above.

The top of the Redwall Limestone is marked by erosion similar to that of many limestone areas today; its surface is extremely irregular, with many old solution caves, sinkholes, and underground channels, a type of topography known as karst. The karst surface gives us a clue to the climate here in late Mississippian time, for such surfaces are known today in Kentucky, Puerto Rico, Yugoslavia, and other limestone-surfaced regions with

LIFE CHANGES

Between Precambrian and Cambrian strata in Grand Canyon, an astonishing change is evident. A look at the Precambrian strata at first glance shows very little in the way of fossils, just some stromatolites: strange low mounds and clusters of stumpy, domed columns formed by mats of primitive algae that trapped and bound calcium carbonate into more or less concentric layers. In contrast, the Cambrian rocks—particularly the Bright Angel Shale—are riddled with signs of life. Trails and burrows, as well as abundant fossil trilobites, brachiopods, bryozoans, and mollusks, illustrate an abundance and diversity of life-forms dwelling in the Cambrian sea.

The seeming burst of life, the sudden diversification, is found across the Precambrian-Cambrian boundary all over the world. But paleontologists suspect that life in Precambrian time may have been more varied than it seems. Precambrian sedimentary strata are restricted in geographical extent, life may not always have existed in sedimentary environments conducive to preservation, and many forms lacked the easily preserved shells of their successors.

Precambrian fossils in Grand Canyon don't yield many clues. The fossils are not easy to get to or easy to see. They include the stromatolites mentioned above, some of whose descendants today live in quiet, shallow waters off western Australia and elsewhere. There are also many small, wrinkled, circular fossils up to about a tenth of an inch (2 to 3 millimeters) in diameter, thought to have been spherical in life—probably some sort of algal cyst. Today many algae make such cysts—durable, long-lasting capsules that enable them to survive periods of drought or other stress. Some microscopic teardrop- or vase-shaped fossils, abundant in some Precambrian layers, resemble modern single-celled protozoa. A few strange tracks and burrows and tiny filamentous plant fossils are known as well.

But where or how life developed, and how it branched out into the multitude of forms evident in Cambrian strata, remain mysteries that the Grand Canyon Precambrian doesn't solve.

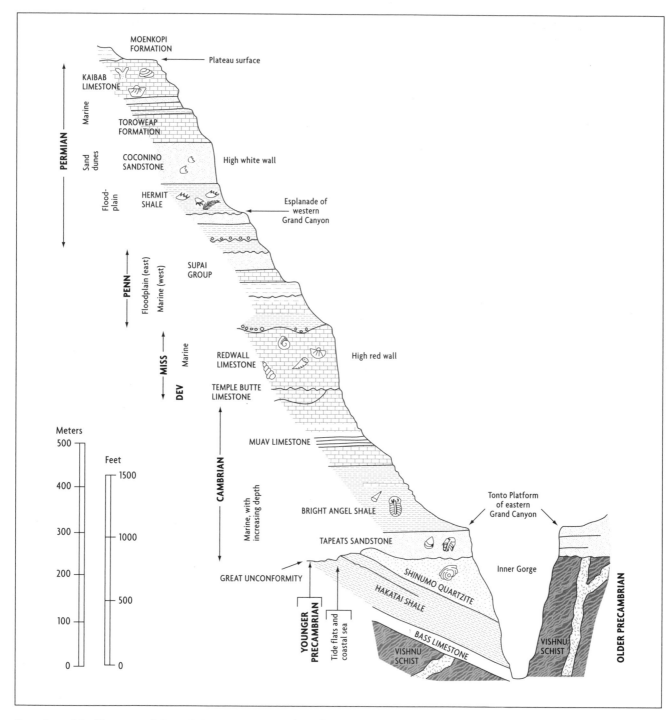

Stratigraphic diagram of Grand Canyon National Park

warm, humid climates. Thin limestone deposited after the karst erosion fills in many of the irregularities in the top of the Redwall Limestone.

During Pennsylvanian and Permian time this region saw more sedimentary deposits. The Supai Group (made up of four formations separated by unconformities) and the Hermit Shale above it consist of ledge-forming, cross-bedded sandstone and deep red slope-forming siltstone. At the time these formations were deposited, the Ancestral Rocky Mountains rose in Colorado and New Mexico. Interbedded sandstone

and siltstone of the Supai Group and Hermit Shale reflect both sea-level changes and increased sediments washed from those newly risen highlands. Spread across broad floodplains and deltas, blown at times by the wind, swept at times by shallow seas, the sediments change rapidly both horizontally and vertically. Some of them bear imprints of fernlike leaves or are marked with reptile tracks; others contain marine fossils such as brachiopods. In Grand Canyon, red siltstone, sandstone, and mudstone deposited on shore thicken and become sandier westward, then yield still

Cambrian fossils from western Grand Canyon include these trilobites. Photo courtesy of Northern Arizona University, Cline Collection.

farther west to marine limestones, showing that the sea lay in that direction.

A little later in Permian time, windblown sand swept across the Grand Canyon region, leaving a thick deposit of fine white sandstone marked with long, diagonal cross-bedding—the Coconino Sandstone, the white cliff on the canyon's upper walls. Tracks and trails of lizardlike animals, as well as trails resembling those created by modern millipedes and scorpions, mark some sloping dune surfaces, but no bones or skeletons of such animals have been found.

During the remainder of Permian time, the western sea continued to fluctuate; twice it inundated the Grand Canyon region, depositing first the Toroweap Formation and then the Kaibab Formation. These rocks now form stairlike ledges and slopes on the uppermost 500 feet (150 meters) of the canyon walls. Again, Grand Canyon exposures allow detailed studies. The lowest part of each formation shows the sea's slow advance, with fragments of underlying rock redeposited across an uneven surface. The middle part of each contains limestone with shore-derived sand in eastern Grand Canyon, pure marine limestone in the west. And the upper parts contain layers of reddish silt and beds of gypsum deposited as the sea backed away westward. Each advance and retreat of the sea is marked with numerous smaller advances and retreats.

Warm and shallow, the Toroweap and Kaibab seas teemed with life, and many fossils can be seen in these rocks: trilobites and brachiopods, snails and clams, cephalopods (shell-bearing cousins of the octopus and squid), feathery bryozoans, corals, and, less commonly, shark teeth.

Mesozoic Era. No Mesozoic rocks are present at Grand Canyon now. At one time, however, Mesozoic strata, deposited above sea level or in a broad interior seaway, covered all of this area. What the Grand Canyon has lost in Mesozoic sediments can still be found in other national parks and monuments described in this book.

At the end of Mesozoic time, with the lifting of the continent and building of the Rocky Mountains farther east, the sea drained away from the continent's interior. Since then the story has been mostly one of uplift and erosion.

Cenozoic Era. Late in Mesozoic time and early in Cenozoic time, ancient Precambrian faults trending north-south were reactivated as the Plateau and surrounding areas began to shift and rise. Paleozoic and Mesozoic sedimentary strata draped across the faults in a series of monoclines (see sidebar on page 112). Mesozoic sediments gradually eroded away, laying bare the more resistant Paleozoic strata below. Scattered patches of stream gravels on top of the Paleozoic sediments indicate that there were highlands to the south and southwest.

Later in Cenozoic time, areas west and south of the Plateau region faulted and pulled apart, numerous basins collapsed between intervening ranges, and today's Basin and Range region became established. The Colorado Plateau remained distinct, a raft in a stormy sea, and began to rise. New faults formed, some not quite parallel to the old Precambrian faults, one of

At Toroweap Point in western Grand Canyon, lava flows repeatedly cascaded into Grand Canyon, damming the river. In this area, erosion of Hermit Shale forms the wide bench of the Esplanade. Tad Nichols photo.

them shaping the sharp west-facing scarp of the Grand Wash Cliffs, here the western boundary of the Colorado Plateau.

This brings us back to our original question: Why is Grand Canyon here, cutting so boldly across the Kaibab Uplift? Geologists find clues in scattered patches of gravel, in lava flows and dikes, in erosional patterns of neighboring areas, in deposits far downstream, and in their knowledge of stream behavior and cliff retreat. A possible sequence of events can be reconstructed from these clues:

- Fault movements early in Cenozoic time reestablished many Precambrian faults in the Grand Canyon area, divided the various plateaus of this region, and domed up the broad anticline of the Kaibab Arch. For some time, streams from this area drained northward.

- By Miocene time, rivers draining the Rocky Mountains of Wyoming and Colorado flowed into large landlocked basins, dumping unbelievable quantities of sediment into Utah, Arizona, and Nevada. In doing so they effectively destroyed their own traces, so their precise routes can no longer be determined.

- Short, steep streams along the western edge of the Plateau deposited locally derived debris close to the cliffs themselves. There may have been a shallow lake west of the cliffs, but no evidence exists of a major river in this area.

- Basin and Range extension, or pulling apart, started around 18 million years ago, dropping the area to the west relative to the Colorado Plateau. Streams along the west edge of the Plateau cut headward with new vigor, dumping debris into basins to the west.

- Around 6 to 10 million years ago, well to the south, between deep faults, a major rift developed, opening a long, narrow seaway that is now the Gulf of California, which for a time extended almost to the present Hoover Dam, far north of its present limit. At nearly the same time, the west side of the Plateau may have sagged a little. Around 5.5 million years ago, a stream that would eventually become the Lower Colorado began to flow south into the new gulf, filling its northern end with estuary deposits. Gradually, other streams joined it, narrowing the divide between their headwaters and the basins where

DOES DAMMING DEEPEN?

Water cascading over a dam tumbles with tremendous force, scouring a churning, often whirling plunge pool below the dam, deepening and widening its channel, ultimately destroying the dam. Man-made dams sidetrack the water for generating electricity, irrigating farmlands, or serving the water needs of nearby or distant population centers. But the ponds or lakes behind natural dams eventually seep through the dam or overtop it, in either case ultimately destroying it.

The Grand Canyon has seen many natural dams come and go. Rockslide dams are easily taken care of by the force of the river. But lava dams are another thing, for lava is hard and tenacious and tough—an impenetrable barrier that can withstand erosion even as adjacent canyon walls erode away. Damming the canyon has long been a favorite sport of volcanoes in western Grand Canyon, where lava has from time to time risen through deep

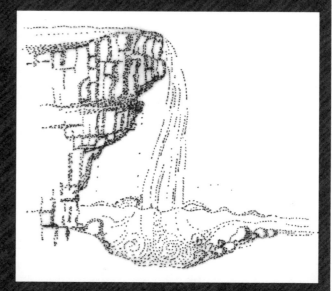

faults, establishing, for instance, volcanic Mount Trumbull, the much smaller cinder cone of Vulcans Throne, and the lava cascades that plunge into the Inner Gorge below them—all features related to the Toroweap Fault.

Lava has dammed the Colorado River in western Grand Canyon many times in the last few million years, creating lakes that extended upstream for many miles, at least one of them as far as the Utah border. Several of the largest lava dams probably held up the Colorado River's flow for years until the lake level rose high enough to overtop the dam, allowing it to resume its downstream journey. Finally cascading over the top of a dam, the river tumbled to the canyon floor, excavating a plunge pool at the foot of the flow. Gradually undermining the lava dam, it thus also deepened the canyon, so that even when the dam finally gave way the river still tumbled in a waterfall, a powerful erosive force that gradually worked its way upstream, eroding the rocks over which it fell, deepening the Grand Canyon, etching out its Inner Gorge.

So though at first it seems contrary to reason, we can interpret these lava flows as a powerhouse for erosion—first damming the river, and then deepening its canyon. This process, repeated many times, could have enabled the Colorado to excavate the great canyon that carves with such impunity through the Kaibab Uplift.

the river or rivers from the Rocky Mountains drained.

- Meanwhile, streams from highlands in central Arizona flowed northward, across the region that is now western Grand Canyon. They may have been tributaries of a larger river system.

- Finally—and possibly fairly suddenly—the divide was breached, the southbound river "captured" the drainage from the Rocky Mountains, and established a direct route to the sea. The geologic "moment" of capture can be recognized in the sudden appearance, in gravel deposits near the Lower Colorado, of tiny fossils characteristic of Cretaceous rocks east of Grand Canyon. The exact way the drainage systems joined to become the Colorado—whether by headward erosion or perhaps by a river or lake overtopping a natural dam—is unknown. How its crooked route over the Kaibab Arch became established is also unknown. That part of the river may have shifted sideways, migrating toward a retreating cliff of Mesozoic rock that covered parts of the Plateau at that time. Erosion has destroyed the evidence.

- Once the river was confined to a narrow trench, downcutting became quite rapid, as all its erosive force was concentrated along a single channel, leaving the work of widening the upper canyon walls to sudden storms, wind, and the force of gravity. With its gradient greatly steepened by the more direct route to the sea, the river used its heavy burden of rock and sand to hammer and pound downward through Mesozoic and Paleozoic strata and into hard Precambrian rocks below. In less than 5 million years, Grand Canyon as we know it took shape.

- In western Grand Canyon, about 1.2 million years ago, volcanoes sent lava flows cascading into the canyon. They show us that by then the canyon was only 50 feet (15 meters) less deep than it is now. So most of the carving of western Grand Canyon took place before then. Dikes that formed as recently as 780,000 years ago show that parts of the Inner Gorge must be younger than that.

- Patches of lava high on Grand Canyon's walls reveal that during its history many lava flows have poured into the canyon. At least twelve times during the last million years, lava flows have dammed the Colorado River, creating lakes that backed up as far as Utah. As the succession of lakes filled and overflowed, waterfalls cascading into turbulent pools undermined the lava dams that created them, then worked their way upstream, helping to carve the canyon (see sidebar).

- Active downward cutting has now slowed as upstream dams regulate water flow and remove the rock debris that the river needs for cutting tools.

BEHIND THE SCENES

Most of the following areas are in eastern Grand Canyon, the part of the national park that sees the most visitors. The remote Toroweap region of western Grand Canyon, though off the beaten track, is included because of its volcanic features and views of the Esplanade.

Bright Angel Point (North Rim). This narrow point, below Grand Canyon Lodge, offers an excellent view down Bright Angel Canyon—deep and straight because it follows the straight line of the Bright Angel Fault.

Partway along this canyon a spot of bright orange marks the Hakatai Shale, part of the Precambrian sedimentary sequence. With other Precambrian sedimentary rocks, it forms a wedge beveled at the top by the Great Unconformity.

The Tapeats Sandstone and younger Paleozoic layers lie above the unconformity. From bottom to top the most prominent layers are the Bright Angel Shale, Muav Limestone, Redwall Limestone, Supai Group, Hermit Shale, Coconino Sandstone, Toroweap Formation, and, finally, the Kaibab Formation at the canyon rim. All these rocks can be identified by referring to the stratigraphic diagram.

Bright Angel Trail (South Rim). Zigzagging steeply, the Bright Angel Trail keeps close to the line of the Bright Angel Fault. Notice the difference between the height of the rim on either side of the fault, where offset rock layers furnish a handy break in the Coconino Sandstone and Redwall Limestone ramparts. The trail follows the fault almost all the way to

Garden Creek in the foreground and Bright Angel Creek in the distance mark the line of the Bright Angel Fault. Indian Gardens, in a clump of cottonwood trees (at arrow), is located where springs rise along the fault. Ray Strauss photo.

the Colorado River, dropping through the entire layer-cake sequence of sedimentary rocks. Trailside signs name the formations. Look in the Kaibab and Toroweap Formations for fossils (no collecting!), in the Coconino Sandstone for dune-style cross-bedding and animal tracks, and in the Hermit and Supai redbeds for ripple marks, leaf impressions, and pebbly layers that establish periods of erosion. The top of the Redwall Limestone shows evidence of solution caverns and rough topography developed before higher layers were deposited.

Not far below the Redwall cliff the trail comes to Indian Gardens, where springs issue from Cambrian rocks to nourish a grove of cottonwood trees. Cambrian Bright Angel Shale borders the trail across the Tonto Platform. At the base of this shale the Tapeats Sandstone forms the ledgy edge of the Inner Gorge. Below the Tapeats is the Great Unconformity, and below that, on the right, the Vishnu Metamorphic Complex. To the left, light-colored Precambrian granite replaces the Vishnu Metamorphic Complex.

Across the river, west of Bright Angel Creek, Precambrian sedimentary rocks lie almost horizontally on top of the Vishnu Metamorphic Complex. Rocks exposed along this part of the Inner Gorge are quite severely faulted, with many small faults cutting across the Bright Angel Fault. Many of the rock layers are sharply bent by drag along the faults.

Cape Royal (North Rim). Cape Royal looks out over a tumbled and jumbled sea of rocks that doesn't follow the Grand Canyon pattern of flat rock layers lying neatly one on another. There are three reasons for this: most of the rocks between Cape Royal and the river are Precambrian sedimentary rocks and lava flows, in wedges tilted in Precambrian time; the Paleozoic rocks themselves curve over the East Kaibab Monocline, the edge of the Kaibab Plateau; and many faults complicate the geologic picture. Some cut and displace only Precambrian rocks; others displace both Precambrian and Paleozoic rocks.

The most easily recognized Precambrian units are bright orange Hakatai Shale and dark Cardenas Lava, evidence of Precambrian volcanism. Tan surfaces near the river are Dox Sandstone. By examining the scene carefully (with binoculars if you have them), you can identify these tilted, beveled Precambrian units and the Paleozoic strata lying horizontally across the bevel.

To the southeast, on the other side of the canyon, a red-sloped butte known as Cedar Mountain rises above the surface of the Marble Platform. It is a remnant of Mesozoic sedimentary rocks that once blanketed this region. Farther north, the Little Colorado River, with an impressive gorge of its own, enters the main Colorado River.

Colorado River. Before the Glen Canyon Dam was built, the Colorado in flood carried downstream an average half-million tons of sand, silt, and gravel per day, and in addition rolled and bounced many large cobbles and boulders along its bed. The muffled roar of rock hitting rock could often be heard from the rim. Descending 2200 feet (730 meters) in 277 miles (439 kilometers), the river at flood stage cleared from its channel much of the bouldery rubble brought in by tributary streams.

Now when such coarse debris is brought to the main channel, the Colorado foams with rapids but can't clear away as much of the rock debris. Without its former "spring cleaning," bouldery deltas build outward from the mouths of large side canyons, constricting the Colorado's channel and enlarging rapids there.

Along the river, rocks are polished and fluted from long abrasion by mud, sand, and gravel. But here, too, we see a slowdown in erosion as the Glen Canyon Dam holds back the river's natural tools. Without boulders for pounding and gravel for scouring and polishing, the river has little cutting power.

River runners see many fine examples of the river's past work. Float trips begin at Lees Ferry at the upper end of Marble Canyon, at the top of the Kaibab Limestone, and descend gradually through the same rock sequence that occurs on the walls of Grand Canyon. Geologic guidebooks for the river run are listed in the references.

Desert View (South Rim). From this viewpoint, one of the most fascinating in the park, you can see the river below its junction with the Little Colorado, which comes in from the south through a deep, narrow, sheer-walled chasm. To the east is Cedar Mountain, a remnant of Triassic rocks left on the otherwise clean-swept Kaibab Limestone of the Marble Platform. Farther away, dim with distance, are Jurassic rocks of Echo Cliffs.

The Kaibab Limestone of the Coconino Plateau bends down sharply along the East Kaibab Monocline—almost at your feet—and then levels off again 3000 feet (1000 meters) below as the surface of the Marble Platform. If you came to Grand Canyon from Cameron, you climbed the East Kaibab Monocline just before entering the park.

In the canyon depths, tilted orange, tan, and gray-green rock layers are Precambrian sedimentary strata that survived the long erosion at the end of Precambrian time. The Great Unconformity cuts across them horizontally, and Paleozoic rocks lie across the truncation.

Looking west down Grand Canyon, notice that the North Rim is higher and farther from the river than the South Rim. Before the canyon was here, the Kaibab and Coconino Plateaus together made up a single broad dome, the Kaibab Arch or Uplift, whose highest point was on the Kaibab Plateau, well north of the present canyon. From that high point, the surface of the plateau (as well as its rock layers) slopes gently southward. Because of this, streams north of the canyon flow toward the river, eroding tributary canyons as they go. By the same token, drainage south of the South Rim

The view northeastward from Desert View shows east-dipping rocks along the East Kaibab Monocline (left), the wide valley carved in faulted Precambrian sedimentary rocks, and the narrow slot of the Little Colorado River's canyon (upper right). In the distance are Mesozoic rocks of Echo Cliffs. Tad Nichols photo.

flows away from the canyon, and takes much less part in the canyon-widening process.

Kaibab Trail (North Rim). This trail descends to the river via Roaring Springs and Bright Angel Canyons, both established along major faults. Paleozoic strata can be identified from the stratigraphic diagram.

Roaring Springs contributes to the water supply for Grand Canyon Village on the South Rim. Water that bursts from these cliffside springs comes from rain and snowmelt that sink into the Kaibab Plateau and flow through a network of underground solution channels in Paleozoic limestones. Reaching impermeable layers of shale, the water moves sideways through gently south-dipping strata until it intercepts the canyon walls.

Because of rapid erosion along Roaring Springs and Bright Angel Faults, there is no broad Tonto Platform here. The trail drops rapidly through all the Paleozoic strata. Below the easily recognized, slabby Tapeats Sandstone it reaches the Great Unconformity, the line of contact with Precambrian rocks. If you place your hand across this contact, your fingers span 600 million years of Earth history.

The uppermost Precambrian rocks are sedimentary and volcanic—Dox Sandstone, Shinumo Quartzite, bright orange-red Hakatai Shale, and Bass Limestone, which you will encounter in this youngest-to-oldest order. At Ribbon Falls, calcium carbonate–laden water plunges over part of a dark diabase sill, depositing its limy load to form a travertine dome. The sill borders the trail for some distance. Farther downstream the trail enters yet older rock, the Vishnu Metamorphic Complex, recognizable by its dark color, shiny crystals, and generally vertical texture. This is the rock of the Inner Gorge, and Bright Angel Creek flows in its own inner gorge most of the way to the river.

The climate change along the trail is pronounced, from a cool Canadian-style forest on the rim to a Sonoran Desert climate at the river. The average temperature difference between rim and river is about 35 degrees Fahrenheit (18 degrees Celsius), making 1 vertical mile (1.6 kilometers) the equivalent of 1200 horizontal miles (2000 kilometers).

Kaibab Trail (South Rim). Descending from Yaki Point, this trail switchbacks through the Kaibab and

Toroweap Formations and the Coconino Sandstone. Many details of these rocks are apparent along the trail: fossils and hard chert nodules in the Kaibab Limestone; red- and yellow-tinted near-shore sandstone interlayered with limestone in the Toroweap Formation; pronounced cross-bedding and wind-ripple marks in the Coconino Sandstone. Notice the fine, even-grained sand weathered from the Coconino Sandstone—typical dune sand.

Below the cliff the trail crosses redbeds of the Hermit Formation and the Supai Group, units that make up red-brown ledges and shelves; as they wash downslope they paint the great Redwall Limestone cliff to match. The karst erosion surface at the top of the Redwall Limestone is rough and irregular, with solution cavities, broken rock, and red-tinged soil similar to limestone-derived soil of the tropics today.

Lees Ferry and Navajo Bridge. Lees Ferry is the launching site for Grand Canyon float trips. The river here is wide and slow. Cream-colored rock ledges near the river belong to the Kaibab Formation, very sandy here near the shore of the Permian sea in which it was deposited. Above Lees Ferry the lowest part of the Vermilion Cliffs is formed of Triassic strata: the dark red Moenkopi Formation at the base, and the variously colored Chinle Formation just above. Higher cliffs expose Jurassic sandstone with dune-type cross-bedding and blue-black desert varnish apparent on many rock faces.

Pima Point (South Rim). Near the west end of West Rim Drive, Pima Point offers a good view westward into less accessible parts of Grand Canyon. Northwest across the canyon is Point Sublime, and to the northeast, about 8 miles (12 kilometers) away, Tiyo Point.

Rock units near and below Pima Point are virtually the same as those elsewhere in eastern Grand Canyon. The Kaibab Formation at the rim contains large Permian brachiopods and knobby masses of chert. Cross-bedding in the Coconino Sandstone shows up particularly well in cliffs just west of Pima Point.

Toroweap Point (North Rim). In western Grand Canyon the pattern of slopes and cliffs changes. The Esplanade, surfaced with soft red sandstone of the Supai Group, replaces the Tonto Platform as the widest shelf in the canyon. A thousand nearly vertical feet (300 meters) below, the Colorado River flows in an inner canyon walled with Redwall Limestone. Cambrian and Precambrian rocks are deep beneath the river.

About a million years ago, great Niagaras of lava cascaded over the canyon's rim and into the river here. Coming from nearby Mount Trumbull, lava streams repeatedly dammed the Colorado River, backing it up into long, narrow lakes. And each time the mighty river eventually cut through the lava dam. White splotches of lake-deposited clay still mark the canyon wall. A younger cinder cone, Vulcans Throne, is right at the brink of the canyon. Eruptions occurred down in the canyon as well as on the rim.

Prospect Creek's steep-walled canyon, entering from the south (left), marks the position of the Toroweap Fault, as does the lava cascade shown on page 103. Rocky flood-built fans like that at Prospect Creek constrict the Colorado River, forming major rapids. Ray Strauss photo.

These volcanic features lie along the Toroweap Fault, which cuts across the canyon here. Toroweap Point is east of the fault, on its upward-moving side. Across the canyon the line of the fault and displacement of the layered rocks are quite apparent. Toroweap Fault extends northward beyond the Utah border, becoming the Sevier Fault.

Yavapai Point (South Rim). Here Grand Canyon is about 9 miles (15 kilometers) from rim to rim. From the museum balcony, Phantom Ranch can be seen in the green patch near the mouth of the long, straight gash of Bright Angel Canyon. With binoculars one can often pick out hikers and mule trains on slender threads of trails that cross the Tonto Platform 4000 feet (1200 meters) below.

Paleozoic sedimentary rocks of the upper canyon walls display their characteristic patterns, weathering into cliffs, ledges, shelves, and slopes, all the effects of differential erosion of resistant and less resistant rock layers. In general, limestone and sandstone form cliffs and ledges, siltstone and shale form slopes. Dark metamorphic rock—the Vishnu Metamorphic Complex—makes up the highest cliffs of the canyon, the walls of the Inner Gorge.

West of Bright Angel Creek, Precambrian sedimentary and volcanic rocks appear beside the river. The eye-catching orange-red Hakatai Shale and the dark Cardenas Lavas are the most easily identified.

Exhibits here explain some of the canyon's other features.

OTHER READING

Belknap, Buzz, 1990. *Grand Canyon River Guide.* Westwater Books.

Beus, Stanley, and Michael Moreles, eds., 2002. *Grand Canyon Geology.* Oxford University Press.

Breed, William, et al., 1996. *Geologic Map of the Eastern Part of Grand Canyon National Park, Arizona.* Grand Canyon Natural History Association.

Elston, Donald Parker, et al., 1989. *Geology of Grand Canyon, Northern Arizona, with Colorado River Guides: Lees Ferry to Pierce Ferry, Arizona.* American Geophysical Union Field Trip Guide Book, T115/315.

Hamblin, W. Kenneth, and J. Keith Rigby, 1968, 1982. *Guidebook to the Colorado River, Parts 1 and 2.* Brigham Young University Department of Geology. This classic is out of print but available on the Internet.

Price, L. Greer, and Sandra Scott, eds., 1999. *An Introduction to Grand Canyon Geology.* Grand Canyon Natural History Association.

Stevens, Larry, 1998. *Colorado River in Grand Canyon: A Comprehensive Guide to Its Natural and Human History.* Red Lake Books.

GRAND STAIRCASE–ESCALANTE NATIONAL MONUMENT

ESTABLISHED: 1996
SIZE: 2969 square miles (7689 square kilometers)
ELEVATION: 3900 feet (1190 meters) at Lake Powell to 9280 feet (2829 meters) at Canaan Peak
ADDRESS: 190 E. Center, Kanab, Utah 84741

STAR FEATURES

✪ More than 200 million years of geologic history revealed in exceptionally well-exposed, mostly Mesozoic strata bent by a series of monoclines and anticlines that clearly influence today's topography.

✪ Narrow, scenic canyons incised by tributaries of the Colorado River.

✪ Scenic backcountry drives and trails leading to features of geologic interest.

✪ BLM visitor centers in Escalante and Cannonville, and the Paria Contact Station on Highway 89, where maps, information, and descriptive leaflets are available.

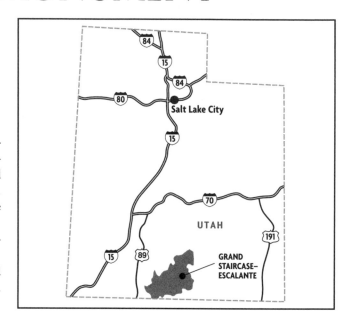

SETTING THE STAGE

Rugged, remote, and colorful, the western part of this national monument was named for a stairlike series of high cliffs and wide, gently sloping benches rising from south to north. East of the Grand Staircase, the Kaiparowits Plateau makes up the central part of the monument. Still farther east is the exciting canyon country of the Escalante River, named for the Spanish father who first explored this area.

The Grand Staircase includes, in ascending order, the Chocolate Cliffs, the Vermilion Cliffs, the White Cliffs, the Gray Cliffs, and, rising north of the national monument, the Pink Cliffs of Bryce Canyon. Between the "risers" of the staircase, the benches or "treads" represent more easily eroded rock layers. With the exception of the Pink Cliffs, all the rocks of the Grand Staircase are Mesozoic in age.

A dramatic monocline, the Cockscomb, defines the eastern edge of the Grand Staircase section of the monument. There, the layers of sedimentary rock curve down sharply toward the east, draped over a deep hidden fault; their eroded edges appear as hogbacks and cuestas below the edge of the Kaiparowits Plateau. Though they level out again beneath the plateau, overlying younger strata surface the plateau itself. Primarily Cretaceous, these younger rocks weather and erode into soft-colored badlands and gentle valleys. In places, canyons cut down through them just enough to reveal the older rock units of the Grand Staircase.

The Kaiparowits Plateau is cut off sharply on the east by the Straight Cliffs, an abrupt escarpment that runs almost without a break for close to 50 miles. The gray Cretaceous rocks that make up these cliffs, from

In the Escalante River section, the Navajo Sandstone erodes into rounded cliffs and surfaces that give the land a hummocky appearance. Lucy Chronic photo.

bottom to top the Tropic Shale and the Straight Cliffs Sandstone, are the equivalent of the Cretaceous rocks that form the lower part of the Gray Cliffs of the Grand Staircase. The light-colored Dakota Sandstone, Jurassic in age, is exposed in hills along the base of the cliffs. Soft Jurassic rocks of the Morrison Formation underlie the valley to the east, and light-colored rocks rising farther east are the Jurassic Navajo Sandstone.

In the easternmost section of the monument, the Escalante River has carved a steep-walled, scenic canyon, cutting down through the Navajo Sandstone. Its tributaries, in turn, have incised a maze of narrow gorges through the same rock and into the red rock below. East of the Escalante River the irregular arc of Circle Cliffs defines the long oval of an anticline whose eastern edge merges with Waterpocket Fold in Capitol Reef National Monument.

GEOLOGIC HISTORY

Paleozoic Era. Cambrian, Devonian, and Mississippian rocks similar to those in Grand Canyon underlie this region but are not exposed in the national monument. Permian limestones, however, come to the surface around the uplift of Buckskin Mountain in the southernmost part of the Grand Staircase area, and in the Circle Cliffs anticline. They show us that early in Permian time, sand and silt washed into this region from mountainous uplifts in western Colorado and northwestern New Mexico, and mingled with limestone deposited in a shallow, fluctuating western sea. Sedimentary rocks that accumulated on shore contain, outside the monument, impressions of fern-like leaves, fossil insects, and amphibian and reptile tracks.

The shallow sea of Permian time extended in an arc across what is now Nevada and the western part of Utah. Twice advancing and retreating, the sea left thin layers of fossil-bearing limestone sandwiched between sandier, siltier shore deposits, the two "sandwiches" making up the Toroweap and Kaibab Formations. Both units contain bands of gypsum and chert typical of near-shore strata. Fossils in the limestones represent environments different from those indicated by the same formations in the Grand Canyon, as we might expect along the shore of the ancient continent.

By mid-Permian time, this region again rose above sea level, and no further sediments accumulated until well into the Mesozoic Era.

Mesozoic Era. The deep red siltstone, sandstone, and shale of the Triassic Moenkopi Formation were deposited on a broad, westward-sloping coastal plain, with the sea not far to the west. Most of the sediments were derived from the still-eroding Ancestral Rocky Mountains in Colorado and other highlands to the south, where ancient Precambrian rocks contributed iron-rich minerals that gave these rocks their rich red color. Deposited by slow-flowing streams and rivers, the sediments in places filled shallow valleys and channels eroded into the Kaibab Limestone surface. The top of the Moenkopi Formation, together with a hard conglomerate from the Chinle Formation, form the Chocolate Cliffs of the Grand Staircase.

By the time the Chinle Formation was deposited, the land had risen a little higher, and was cut off from

Goblins in Devils Garden form in the Entrada Sandstone. Subtle differences in texture and cement contribute to their strange shapes. Lucy Chronic photo.

the sea to the west. Apparently deposited by broad, meandering streams and in shallow, perhaps seasonal lakes, it carries evidence of abundant and varied life: petrified logs up to 90 feet long, fossils of other plants, fossil fishes, amphibians, and reptiles, and tracks of early dinosaurs. Clays derived from volcanic ash point to volcanoes somewhere to windward—probably in what are now California and southern Arizona.

After another interlude of erosion, Jurassic sand dunes swept across this region, creating the Wingate Sandstone, well exposed in Circle Cliffs. Dune-formed sandstones seem to have alternated with river flood-plain deposits through much of Jurassic time, with the addition of occasional lake deposits and layers of clay derived from volcanic ash. The Wingate Sandstone, marked with the long diagonal striations that indicate its sand dune origin, changes westward into the Moenave Formation, red, silty river and lake deposits much like the Triassic rocks described above. Fossils are sparse in both these formations, but dinosaur tracks and a few fish remains give us a hint of the lowland and lake environments in which they formed.

More river and lake deposits, the sandstone and silt-stone layers of the Kayenta Formation, lie above the Wingate Sandstone. They are also Jurassic in age, and they, too, contain dinosaur tracks, fragments of petrified wood, and bones of vertebrate animals. The top of the Kayenta Formation intertongues with the overlying pale buff Navajo Sandstone—the White Cliffs of the Grand Staircase, the formation whose cliffs and rim rocks contribute so much to the striking beauty of the

canyons of the Escalante River and its tributaries. With uniform, rounded grains and distinctive cross-bedding, the Navajo is clearly another wind-deposited sandstone. It represents an ancient desert, an arid, dune-covered expanse rivaling the present Sahara. It thickens markedly westward to stand in the high cliffs of Zion National Park, where it varies in color from white to pink. Where did all this sand come from? We don't really know. Deserts like this develop today between 15 and 35 degrees from the equator.

Jurassic sedimentary rocks younger than the Navajo Sandstone are present in many places in the national monument. Commonly lumped together as the San Rafael Group, they include, again, floodplain and dune deposits. The uppermost Jurassic rock unit here, the Morrison Formation, was deposited by slow-moving, winding streams, with occasional input of volcanic ash or development of sand dunes. Its soft, fine, clayey layers and pastel green and purple hues, with layers of yellowish sandstone—the colors due to nonoxidized iron minerals that suggest a marshy or swampy environment—make it easy to recognize. Though the Morrison Formation is famous elsewhere for dinosaur fossils, none have been found in Grand Staircase–Escalante National Monument. Here, however, a few unidentifiable reptile remains have been located.

A long episode of erosion, spanning half of Cretaceous time, created an unconformity that divides Jurassic from Cretaceous rocks. As the land began to subside in mid-Cretaceous time, an inland sea spread across the center of the continent. Developing mountains in

MAKING MONOCLINES

The steplike bends of many monoclines mark the surface of the southern Colorado Plateau: the East Kaibab Monocline, the Cockscomb, Waterpocket Fold, and others. Exposures of Precambrian rocks in the depths of Grand Canyon reveal their structure: basically flat-lying sedimentary rocks draped across ancient Precambrian faults. Some of the monoclines, particularly the Cockscomb and Comb Ridge, are characterized by sharp-edged hogbacks produced by erosion of the sloping sedimentary layers.

Most of the monoclines run north-south, a trend inherited from Precambrian time. Their usual up-on-the-west, down-on-the-east character, however, may be the opposite of the original Precambrian movement—as is true in the two faults exposed in Grand Canyon.

The monoclines bend Paleozoic and Mesozoic rocks but not younger Tertiary ones, so they are known to have developed in very late Cretaceous or very early Tertiary time. At this time, subduction of the Farallon tectonic plate beneath the western part of North America caused compression across this region and resulted in uplift of the Colorado Plateau and of the Rocky Mountains to the east. However, smaller folds in this area, as well as faults, fracture (joint) patterns, and many small anticlines and synclines, follow a different trend; they run northwest-southeast. Strains in the crust in Precambrian time apparently differed from those involved in the compressional force of the Laramide event. Fracturing also occurred with additional Miocene-Pliocene regional uplift of the Colorado Plateau.

Monoclines were first recognized and defined by John Wesley Powell in 1873, on the basis of his observations in this region, which is therefore considered the "type area" for this kind of geologic structure.

western Utah, created by the Sevier Orogeny, limited westward expansion of this inland sea, as did rising highlands in central Arizona.

The Dakota Formation—a thin but widespread unit—developed as the sea receded across the beveled surface of the Jurassic rocks. Predominantly sandstone, the Dakota Formation displays some beach or stream cross-bedding; it also includes plant fragments, coal, ripple marks, and other indications of a near-shore environment. A fluctuating advance of the sea is written in its uppermost layers, which contain a few marine fossils.

Above the Dakota Formation lies the soft gray, silty marine Tropic Shale, generally a slope-former, deposited as fine gray mud on the floor of the Cretaceous sea. Localized thickening of the Tropic Shale and the rocks that overlie it show that there was some broad, gentle warping of the crust in southern Utah, creating north-south anticlinal uplifts and synclinal basins in the sea's floor. The upper part of the Tropic Shale probably represents maximum depth of the sea in this area.

The retreat of the sea is seen in the Straight Cliffs Sandstone, which overlies the Tropic Shale and contains marine fossils near its base but includes stream-formed cross-bedding, petrified wood, and bits of fossilized bone in its higher layers.

With continued growth of mountains in the west, additional terrestrially deposited sediments reached into this region. Uppermost Cretaceous deposits are visible rising as hills and cliffs north of the monument, though they are not present in the monument itself.

Cenozoic Era. Upward arching of the continent and the building of mountain ranges, including the Rocky Mountains, characterize the Cenozoic Era. At the beginning of the Cenozoic, ancient basement faults reactivated, resulting in the monoclines that drape down to the east here and elsewhere (see sidebar). Paleocene lake and river deposits of the Claron Formation, which form the Pink Cliffs of Bryce Canyon and the Aquarius Plateau, were deposited and can be seen in the distance from several viewpoints in the monument. Further uplift of the Plateau region in Pliocene-Miocene time, and the eventual reorganization of drainages into the Colorado River system, resulted in the removal of vast amounts of overlying sediment and the downcutting of the deep canyons of the Plateau.

BEHIND THE SCENES

The Blues. Well up the mountain slope east of Henrieville, a thick, gray, muddy sandstone that is part of the Kaiparowits Formation erodes into a badland area of mounds and hills that appear almost blue. The Kaiparowits is at its thickest here—almost 3000 feet (915 meters)—and contains plant fossils, freshwater snails, and vertebrate fossils such as small mammals, crocodiles, turtles, and dinosaur fragments. Powell Point, the sharp, pointed cliff above the Blues, is topped

Jointing controls this unusual biscuit-shaped pattern in the Navajo Sandstone. Halka Chronic photo.

with Paleocene lake deposits of the Claron Formation, the same formation that appears in the distance in the Pink Cliffs of Bryce Canyon National Park.

Burr Trail. This paved road traverses the eastern section of the monument from the town of Boulder to Capitol Reef National Park. It winds at first across stream gravels deposited by Boulder Creek, but soon passes among rounded exposures of Navajo Sandstone. Watch for "checkerboard" patterns on the rock, where small gullies along vertical joints crisscross the wind-deposited cross-beds of the formation. This is an ideal area in which to observe the broad cross-bedding characteristic of dune deposits. Farther east, older rocks uplifted by the Circle Cliffs anticline appear in vertical red cliffs of a narrow canyon walled with Wingate Sandstone. The Kayenta Formation is visible at the top of the cliffs.

The Wingate Sandstone also makes up the 300-foot (91-meter) Circle Cliffs, best seen from a side trip along the Wolverine Loop. Within the Circle Cliffs anticline, rocks as old as the Permian Kaibab Formation come to the surface. Tar deposits and remnants of a breached oil reservoir, as well as petrified logs, can be seen along this loop. Since oil floats on water, it tends to rise through sandstone or other porous rock. In anticlines, it often pools below layers of impervious shale, so anticlines are likely places for petroleum exploration. Oil-bearing rocks in this anticline were breached long ago by erosion, and it now contains no commercial-quality oil. Another anticline west of Escalante does have oil wells that reach oil from Permian rocks below the surface.

From high points along the Burr Trail, the Henry Mountains are in view to the east. They consist of a cluster of small igneous intrusions that, in Tertiary time, domed up overlying sedimentary rocks. Except around the base of the mountains, the sedimentary layers have eroded away, leaving only the hard igneous core of the range. The more distant La Sal and Abajo Mountains formed the same way.

The Burr Trail continues into Capitol Reef National Park, where it descends in steep, narrow switchbacks down the dramatic monocline of Waterpocket Fold.

The Cockscomb. Along the monocline that borders the east side of Buckskin Mountain, Triassic and Jurassic formations bend sharply, in places even vertically, with resistant layers eroded into hogbacks that rank up along the east flank of the mountain. Notice the many small faults that offset the strata in highway cuts along US Highway 89. Buckskin Mountain is surfaced with Permian Kaibab Limestone, the resistant rock that also tops the Kaibab Plateau and forms the rims of Grand Canyon.

Escalante River and Trails. In the scenic Escalante River section of the national monument, erosion has incised deep canyons into rugged rock benchlands and slickrock surfaces of Navajo Sandstone. Here the Navajo, not tightly cemented, tends to erode roundly, without sharp corners, giving the surface a dunelike, undulating appearance that almost mimics the Jurassic dunes of which the rock is formed. Into this surface the Escalante's many small tributaries have cut canyons, narrow and winding, some of them mere slots a few feet wide. Along the Escalante River and some of these tributaries, the bright red of older rock units is visible at the base of the canyon walls, further dramatizing the scenery. Utah 12 east and west of Escalante winds in and out of several of the narrow gorges, and trails provide access to others.

Grosvenor Arch. This unusual double arch, actually an arch and a window, accessible from Cottonwood Wash Road, owes its irregular shape to the ways that three formations—the Henrieville Sandstone, the Cedar Mountain Formation, and part of the Dakota Formation—erode. Natural arches form as narrow ridges or fins of rock are eroded from both sides by rain, windblown sand, and frost, often where very slight seeps weaken the rock by dissolving away the calcium carbonate that holds sand grains together. Vertical joints in the rock control the position and shape of the arches.

Hole-in-the-Rock Road. The Straight Cliffs control the route of the Hole-in-the-Rock Road. A tall, unbroken escarpment running almost without break for 50 miles, the Straight Cliffs also parallel the joint trend and the trend of many monoclines, anticlines, and synclines of this part of the Colorado Plateau, as well as the direction of many stream courses like that of the Escalante River and small tributaries north of Lake Powell. The rocks that form this linear cliff—the Dakota Sandstone at its base, the Tropic Shale and the Straight Cliffs Formation above—represent comings and goings of the shallow Cretaceous sea. The Tropic Shale is marine; the Straight Cliffs Formation contains

Overlying sediments filled an earthcrack in still-soft muds below, creating a clastic dike. Lucy Chronic photo.

river sands and muds as well as sediments deposited on beaches or deltas, offshore bars, or other coastal terrain.

Some distance south along Hole-in-the-Rock Road, badlands have developed, with strange mushroom rocks and other erosional figures in the Entrada Sandstone. At Devils Garden, the Entrada Sandstone is almost flat-lying, and has eroded into many short spires, knobby figures, and arches. Such figures are fairly common in outcrops of the Entrada Sandstone, where erosion breaks down and washes away parts of soft shale layers, undermining harder sandstone above.

Paria Movie Set Site. Its colorful jumble of bad-lands and cliffs has made this remote area a favorite backdrop for western movies. Triassic deposits of the Chinle Formation erode into boldly striped mounds, with the Vermilion Cliffs rising red and stark above them. The badlands form because poorly consoli-dated sediments—high in clay, especially clay derived from volcanic ash—erode too quickly for soils to form on them.

OTHER READING

Abbey, Ed, 1971. "Escalante Canyon" in Meyer, Alfred, ed., 1971. *Encountering the Environment.* Van Nostrand Reinhold.

Chesher, Greer, 2000. *Heart of the Desert Wild: Grand Staircase–Escalante National Monument.* Bryce Canyon Natural History Association.

Fleischner, Thomas Lowe, 1999. *Singing Stone: A Natural History of the Escalante Canyons.* University of Utah Press.

Lambrechtse, Rudi, 1985. *Hiking the Escalante.* Wasatch Publishers.

HOVENWEEP AND CANYONS OF THE ANCIENTS NATIONAL MONUMENTS

HOVENWEEP NATIONAL MONUMENT

ESTABLISHED: 1923
SIZE: 1.25 square miles (3.2 square kilometers)
ELEVATION: 5100 to 5980 feet (1554 to 1822 meters)
ADDRESS: McElmo Route, Cortez, Colorado 81321

CANYONS OF THE ANCIENTS NATIONAL MONUMENT

ESTABLISHED: 2000
SIZE: 256 square miles (663 square kilometers)
ELEVATION: 5080 to 6600 feet (1548 to 2000 meters)
ADDRESS: 27501 Highway 184, Dolores, Colorado 81323

STAR FEATURES

- ✪ Nearly horizontal Jurassic and Cretaceous strata of a broad tableland, cut by steep-walled canyons containing hundreds of Anasazi archaeological sites.
- ✪ A visitor center at Hovenweep National Monument, and the Anasazi Heritage Center in Dolores, both offering displays, information, and interpretive presentations.
- ✪ Interpretive walks and brochures at individual ruins and groups of ruins.

SETTING THE STAGE

Hovenweep and Canyons of the Ancients National Monuments lie northwest of Cortez, Colorado, on a broad tableland that slopes gently westward into Utah. Almost featureless when seen from a distance, the tableland is scored by numerous small, hidden canyons, steep-walled gorges that 900 to 1100 years ago sheltered a scattered population of farmers related to the pueblo-builders of Mesa Verde and many other archaeological sites in the Plateau region.

In Hovenweep, just two formations make up the landscape: the Cretaceous Dakota Sandstone at the surface and in cliffs and ledges of the canyon, and below them the Jurassic Morrison Formation, soft mudstone, sandstone, and volcanic ash eroded into slopes and recesses. Fine, even-grained windblown silt—a product of the Pleistocene Ice Age—provides the fertile soil attractive to both prehistoric and modern farmers.

In Canyons of the Ancients National Monument, the Dakota Sandstone and Morrison Formation are also present. Closer to Sleeping Ute Mountain, the canyons are somewhat deeper than at Hovenweep. As a result they reveal the base of the Morrison Formation and, beneath it, two massive Entrada Sandstone cliffs.

GEOLOGIC HISTORY

Paleozoic Era. While no Paleozoic rocks are exposed within these two national monuments, we know from studies of surrounding areas, as well as from

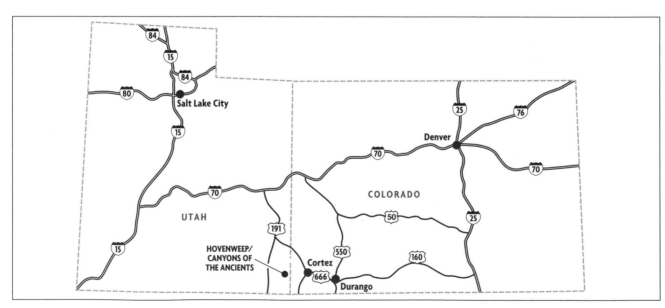

records of oil and gas drilling in the vicinity, that flat-lying layers of early Paleozoic marine sedimentary rocks are overlain by Pennsylvanian and Permian river floodplain and delta sediments that reflect uplift of Uncompahgria, the westernmost range of Colorado's Ancestral Rocky Mountains. We know also that the Pennsylvanian strata contain thick layers of salt deposited in the deepening, partly landlocked basin that lay west of Uncompahgria.

Mesozoic Era. The oldest Mesozoic rock visible here is the Entrada Sandstone, the smooth-surfaced, tan and salmon-colored rock that forms the lowest cliff in Sand Canyon in Canyons of the Ancients National Monument. The Entrada Sandstone bears all the hallmarks of windblown sand, for it formed as Jurassic dunes swept across this area next to an inland sea. Its sand grains are rounded and even in size, and its cliffs are marked with the long, diagonal cross-bedding characteristic of dune deposits. Thin horizontal layers in the sandstone represent flat areas among the dunes, where fine, salty mud accumulated in shallow, short-lived desert lakes, the equivalent of the sabhkas known in the Sahara today.

Silty layers with evaporites overlying the sandstone, forming the slope above the cliff, may also be sabhka deposits. Some of them appear to be shallow near-shore deposits affected by tides, however, with ripple marks formed by wave action. Thinly bedded, smooth-surfaced siltstones may have been deposited in quiet, perhaps deeper water.

Eventually the sea retreated and dunes appeared again, as represented by the second of the cliff-formers in Sand Canyon—another cross-bedded sandstone. This unit may be the uppermost part of the Entrada Sandstone or the lowest part of the overlying Morrison Formation: exact relationships are hard to define in this region where outcrops are so limited, and the wind-deposited sandstone contains no fossils or other clues by which it can be dated.

By late in Jurassic time, the sand dunes were gone, the continental seaway had departed, and rivers swept from western and southern mountains across broad, nearly level plains. These brought fine sand, silt, and mud well loaded with volcanic ash from explosive volcanoes up and down the length of what is now California. The resulting greenish, purplish, and in places yellowish Morrison Formation is widespread and particularly easy to recognize.

By the end of Jurassic time, and with continued growth of uplands in Nevada and California, the scene changed again. Erosion of several hundred to perhaps as much as a thousand feet of sediments took place before the Burro Canyon Formation and Dakota Sandstone were deposited in Cretaceous time. These deposits, visible in Little Ruin Canyon near the Hovenweep Visitor Center, are coarser—brought in by streams and

Horizontal layers, or bedding, in the Dakota Sandstone made it a convenient building material for these towers. Lucy Chronic photo.

rivers flowing northeastward from the mountains. The Burro Canyon Formation, a discontinuous stream or river deposit, is topped by another unconformity. The Dakota Sandstone, generally thought to be a near-shore strandline or beach deposit left behind during the retreat of the Cretaceous inland sea, contains river deposits here as well.

Later Cretaceous deposits, such as the Mancos Shale and Mesaverde Group, once extended across this area but have since eroded away.

Cenozoic Era. In early Tertiary time, just east of these national monuments, a mass of molten rock pushed its way upward and spread outward into surrounding deposits, forming the Sleeping Ute laccolith. The intrusion lifted Cretaceous strata and formed the present Sleeping Ute Mountain. Cretaceous rocks have eroded off the mountain and, by a slow process of cliff retreat, off the land west of the mountain as well, baring the surface of the more resistant Dakota Sandstone. On it, the winds of Pleistocene and Recent times deposited the layer of fine silt that provided the Anasazi and later farmers with fertile soil (see sidebar).

OTHER READING

Thompson, Ian, 1993. *Towers of Hovenweep*. Mesa Verde Museum Association.

WHY SO MANY RUINS?

This is arid country, and yet here on these mesas ancient populations once reached high densities. Why here? Look to the geology for an answer.

Shallow canyons, cut in horizontal layers of sedimentary rock, are invisible from the surface of the plain until you stand next to them. They are warm and sheltered in the winter, especially against the south-facing walls. Building materials are readily available along the eroded edge of the canyons—sandstone slabs and chunks that cave off the steep edges. Seeps and springs provide water that percolates through sandy formations like the Dakota Sandstone, reaches impermeable clay layers below the sandy ones, and flows along them to the surface. And on the surface of the plain, soil derived from Pleistocene and Recent wind deposits provides good farmland.

Builders at Hovenweep took advantage of an irregularly shaped block of sandstone.

At the time of the occupation of these canyons, rainfall and warmth must have been critically balanced for successful farming. Archaeological records show that populations here fluctuated in response to rainfall, temperature and, possibly, marauding bands of enemies. Two main influxes occurred, with two population peaks, but the canyons were deserted by the 1300s.

Today, soils above the canyons are farmed again, with wells providing water adequate for many crops.

MESA VERDE NATIONAL PARK

ESTABLISHED: 1906
SIZE: 81 square miles (211 square kilometers)
ELEVATION: 6025 to 8305 feet (1836 to 2531 meters)
ADDRESS: PO Box 8, Mesa Verde National Park, Colorado 81330-0008

STAR FEATURES

- ✪ A high mesa whose walls portray the encroachment and departure of a Cretaceous sea.
- ✪ Close looks at erosional processes leading to formation of caves in horizontal strata.
- ✪ Views of the Four Corners country to the south, including the famous landmark Ship Rock, the San Juan Basin, and the broad furrow of the San Juan River.
- ✪ Archaeological remains on the mesa surface and in large rock alcoves: masonry villages of Anasazi peoples who inhabited this region from A.D. 600 to 1300.
- ✪ Visitor center, museum, introductory program, two loop drives, guided walks, self-guided trails, and shuttle bus trips to ruins. At most ruins visitors must be accompanied by park interpreters.

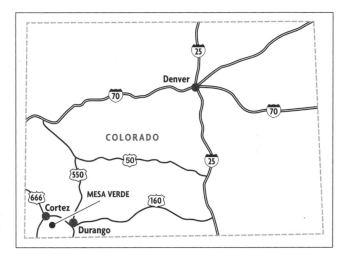

SETTING THE STAGE

Like Chaco Canyon, Mesa Verde National Park, primarily of archaeological interest, also tells a geologic story.

A high, southward-sloping tableland that drops off precipitously in all directions, Mesa Verde is composed almost entirely of Cretaceous rocks. Its steep slopes reveal nearly 2000 feet (600 meters) of gray siltstone: the Mancos Shale, deposited fairly far from shore in a Cretaceous sea. This formation contains fossil marine shellfish as well as occasional bones and teeth of fishes or of large marine reptiles known as plesiosaurs.

Weak and easily eroded, prone to landslides (to the detriment of the park entrance road), the Mancos Shale is capped with much harder rock—sandstones of the Mesaverde Group. This group, once considered a single formation, is now subdivided into three units, which we can look at in the bottom-to-top order in which they were deposited.

The lowest is the Point Lookout Sandstone, the rock that caps the north end of Mesa Verde, including the prominent north-jutting spine of Point Lookout. About 400 feet (120 meters) thick, this unit is a fine-grained, cross-bedded sandstone, largely a beach deposit. It con-

tains fossils of both land plants and marine shellfish.

Above the Point Lookout Sandstone, the Menefee Formation is primarily nonmarine, alternating layers of shale, siltstone, and sandstone, with numerous thin bands of coal. It varies from 340 to 800 feet (100 to 250 meters) in thickness, and appears in canyons of the southern end of Mesa Verde, where it forms slopes below the mesa cap.

Capping the southern part of Mesa Verde and sheltering its famous cliff dwellings, the Cliffhouse Sandstone is massive, fine grained, and cross-bedded—again a beach or shallow marine coastal deposit. It contains a few thin layers of shale and coal, as well as marine fossils.

Younger Cretaceous formations have been stripped away by erosion. The three formations of the Mesaverde Group dip southward at a slightly steeper angle than the gentle slope of the mesa surface. Because of this, the upper two units of the Mesaverde Group—the Menefee Formation and Cliffhouse Sandstone—are eroded from the north end of the mesa. At the south end, the lowest unit—the Point Lookout Sandstone—is below the surface, even below the floors of the deep canyons there. The beveling of these formations is thought to have taken place before Mesa Verde was separated from the San Juan Mountains to the north, when mountain-derived streams carved a broad southern pediment. Some of Mesa Verde's canyons, including Prater and Morefield Canyons, were carved during this time.

In floor plan the mesa is something like a human

Spruce Tree House was built in a deep recess in the Cliffhouse Sandstone. Springs that aided in creating the recess also furnished water for twelfth- and thirteenth-century inhabitants. Halka Chronic photo.

hand, with deep south-draining canyons separating its long fingers. Many of the canyons are parallel, their position controlled by parallel north-south joints in the rock. Along the walls of nearly all the canyons, erosion has undermined the massive rimrock, causing large blocks to fall away. Where surface water percolates downward through the porous sandstone, it meets the impervious shales and moves sideways

Slope-forming Mancos Shale undermines a cliff of Mesaverde Group sandstone that represents the near-shore deposits of a retreating Cretaceous sea. L. C. Huff photo, courtesy of USGS.

until, at the canyon walls, it emerges as springs. Undermining is particularly effective near springs, where, conveniently for Anasazi villagers, large arcuate caves or alcoves had formed as arching sheets of sandstone fell away. Flat-floored alcoves protected by thick ledges of sandstone, provided with building blocks fallen from the ceiling, fed by springs of clear water—what better sites for villages?

Vital to Mesa Verde's prehistoric inhabitants was the fine, even-grained soil covering the surface of the mesa, too fine-grained to have been derived from the Cliffhouse Sandstone that caps the mesa. The soil, primarily wind-deposited Pleistocene silt, a product of the Ice Ages, is fine and easily tilled, and was undoubtedly attractive to the Anasazi farmers who settled here.

Mesa Verde's cliff houses were discovered during the 1880s by members of a government-sponsored geological and geographical survey of western territories. Archaeological work here began in the 1890s, and intensified after the area became a national park in 1906. Early in the 1930s many of the ruins were dated by analyzing the tree rings of logs used for ceiling beams. By 1935, about 4000 sites of prehistoric habitation had been identified—some on the mesa surface, some in caves below the rim.

GEOLOGIC HISTORY

Mesozoic Era. Mesa Verde's geologic story begins with the advance of the Cretaceous sea, the last sea to cover the interior of the continent. Creeping over a land that for all of Triassic and Jurassic time had been above sea level, the sea came from the east and northeast. At first, as the land subsided, the sea swept in fairly rapidly and then retreated, leaving behind the layer of beach and near-shore deposits that became the Dakota Sandstone. The Dakota is a thin but widespread rock that now surfaces the area around the town of Cortez and also appears in hogbacks both east and west of the Rocky Mountains.

With further subsidence, the shoreline crept westward, and the sea in the Mesa Verde area deepened. In it, fine soft muds accumulated, eventually to become the Mancos Shale. Millions of years in the making, this shale is nearly as widespread as the Dakota Sandstone, but a great deal thicker.

When at last the Mancos sea retreated, coarser sediments accumulated along its shores to form the Point Lookout Sandstone. Above this unit the rocks of the Menefee Formation were deposited by meandering rivers and streams and in shallow swamps and marshes. A good deal of plant material also accumulated there, later to be compressed and altered into coal. Plant fossils, including conifers, palms, cypresses, and magnolias, tell us that the climate when the Menefee Formation was deposited was much like that of the Gulf Coast today.

As the land level fluctuated once more, another thick layer of near-shore sand, the Cliffhouse Sandstone, was deposited, the uppermost and youngest layer of the Mesaverde Group. Here again, not all the material that went into this unit was pure sand. In quiet bays and estuaries fine silt and clay also accumulated, forming thin, shaly layers that eventually played a part in undermining the massive cliffs of Cliffhouse Sandstone, creating the great alcoves of Mesa Verde. Fossils in the Cliffhouse Sandstone include clams, ammonites, snails, and shark's teeth.

After accumulation of the Mesa Verde Group, the sea deepened again, leaving fine gray marine muds of the Lewis Shale overlain by the Pictured Cliff Sandstone, both now largely eroded off Mesa Verde. Late in Cretaceous time, as the Rocky Mountains began their upward push, the sea drained away completely.

Cenozoic Era. The rise of the Rockies lasted well into Cenozoic time. In what is now southwestern Colorado, a broad dome, 100 miles across, rose as one of the ranges of the Rockies. Erosion whittled at this uplift, carving down through domed layers of Mesozoic and Paleozoic rocks and into underlying Precambrian rocks. Along the southern side of the dome, where sedimentary rocks dip gently southward, both the Pictured Cliff Sandstone and the Lewis Shale were slowly eroded away.

Starting about 40 million years ago, volcanic activity completely changed the nature of the great dome. Eruptions lasted nearly 30 million years, with incredibly violent outpourings that destroyed whole stratovolcanoes as they collapsed into partly empty magma chambers, only to have new volcanoes build on the ruins of the old, creating the high, rugged range of the San Juan Mountains northeast of Mesa Verde.

Late in the history of this area, the entire region,

Along the northwest edge of Mesa Verde, streams that once drained from the San Juan Mountains to the north have been "beheaded" and deprived of their water source as erosion of the Mancos Valley cut Mesa Verde off from the mountains. Halka Chronic photo.

from New Mexico to Montana, was lifted some 5000 feet, to its present elevations. As always, uplift increased erosion. With new vigor, streams draining the San Juan Mountains cut into surrounding areas, separating Mesa Verde and several similarly isolated mesas from the mountains. As the San Juan River deepened its valley, small tributaries draining the southern part of Mesa Verde eroded headward, cutting the narrow canyons that separate the southern fingers of the mesa. Aided by natural springs, the streams undermined the Cliffhouse Sandstone, creating the big alcoves that later became Anasazi building sites.

Erosion continues here today as soft sediments give way in landslides along the margins of the mesa, and large blocks of sandstone are undermined and tumble to the slopes below. Lichens establish them-selves on rock surfaces, beginning the development of soil. Tree roots pry rocks apart. Rain and snowmelt initiate rivulets that course down steep slopes, carving rills that will eventually turn into gullies that will turn into canyons. The tips of the mesa fingers will be cut off from the main mesa, surviving for a time as buttes and pinnacles, then disappearing altogether.

OTHER READING

Griffitts, Mary O., 1989. *Guide to the Geology of Mesa Verde National Park.* Mesa Verde Museum Association.

Martin, Linda, 2001. *Mesa Verde: The Story Behind the Scenery.* KC Publications.

Wenger, Gilbert O., 1980. *The Story of Mesa Verde National Park.* Mesa Verde Museum Association.

NATURAL BRIDGES NATIONAL MONUMENT

ESTABLISHED: 1908
SIZE: 12 square miles (20 square kilometers)
ELEVATION: 6500 feet (1980 meters) at visitor center
ADDRESS: PO Box 1, Lake Powell, Utah 84533

STAR FEATURES

✪ Three natural bridges, each showing a different stage in bridge development, carved in cross-bedded Permian sandstone.

✪ Numerous other erosional features, as well as examples of soil development in a semiarid climate.

✪ Triassic rocks mainly but not entirely eroded off the mesa surface.

✪ Visitor center, museum, introductory slide show, loop drive to the bridges.

SETTING THE STAGE

In White and Armstrong Canyons, which cut deeply through the horizontal surface of Cedar Mesa, three natural bridges have formed in rocks of the light-colored, cross-bedded Cedar Mesa Sandstone. Natural bridges, in contrast to arches, span the stream courses responsible for their formation (see sidebar). They can develop by several methods. Here, three aspects of the geologic setting have encouraged bridge formation.

• Rock units are horizontal.

• Streams that once wound across the mesa surface have retained their sinuous courses even as they cut deep into the sandstone. Entrenched meanders thus loop around narrow fins of rock extending from the mesa surface.

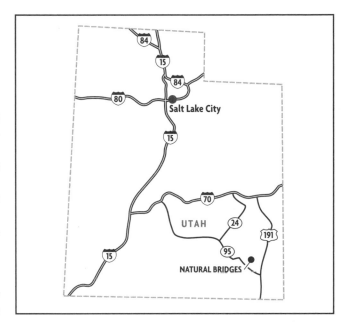

• The Cedar Mesa Sandstone is conducive to arch formation. It is composed of cross-bedded dune sand with fine, evenly sized grains cemented together with calcium carbonate. The cross-bedding is interrupted by horizontal layers of fine siltstone deposited in low areas between the former dunes.

In the semiarid climate of southern Utah, the streams in these canyons, though often dry, occasionally sweep forcibly against the jutting rock fins, grinding out the soft siltstone layers of the interdunes and pounding the sandstone with sand, pebbles, and boulders. Gradually

---

BRIDGES, ARCHES, AND WINDOWS

What is the difference between an arch, a window, and a natural bridge? Geologic features such as these can be likened to their man-made counterparts. Windows are relatively small and high up on the sides of cliffs—or buildings. Natural bridges, like man-made bridges, span stream courses—whether or not the streams have water in them. And arches—which need not be rounded—do not span stream courses. Arches, both natural and man-made, may not be freestanding.

Many bridges, arches, and windows, however, have been misnamed in the past. Some of the "arches" in Arches National Park are technically windows, for instance, and the "double arch" of Grosvenor Arch is an arch with a window high up on its side. Others are correctly named: Angels Window in Grand Canyon is indeed up on a high rock wall; Rainbow Bridge spans what is now known as Bridge Creek; and Delicate Arch, Landscape Arch, and many others in Arches National Park do not span watercourses.

Slender Owachomo Bridge formed at the junction of White and Tuwa Canyons. Halka Chronic photo.

the fins become thinner until finally they break through, leaving natural bridges.

Flowing water seeks the most direct path downslope. Once a bridge has formed, the stream abandons its meander loop and shortcuts under the bridge. As time goes on, the stream further undermines the rock and enlarges the opening below the bridge. With rain and snow, with day-to-night and season-to-season temperature changes, thin spalls of rock flake off from the undersurface of the span. Gradually the massive bridge becomes thinner and thinner until only a narrow arch of rock frames the sky above the stream course.

The three large bridges featured in this national monument represent three stages in the life of such natural bridges. Kachina Bridge exemplifies an early stage. It is thick and bulky, with a relatively small passage below. Sipapu Bridge is thinner, more graceful. The stream is no longer eroding its abutments, but erosion of the bridge itself is continued by rain, wind, and frost. Owachomo Bridge is in a late stage of bridge development. Its span is slender and increasingly fragile. In the end, gravity will become too great a stress for this delicate span, and Owachomo Bridge will fall. As Sipapu and Kachina Bridges similarly become more and more delicate, other fins, pounded by flooding streams, will break through to create new bridges.

The rock in which the bridges took shape—the Permian Cedar Mesa Sandstone—shows many other interesting features, easy to see along the bridge trails. The steep dune-type cross-bedding is interrupted here and there by contorted wiggles that show where sand slumped or avalanched down lee faces of dunes. Even

Natural bridges develop where streams hammer at narrow fins of rock. At this entrenched meander or gooseneck, the stream may eventually carve another bridge. Halka Chronic photo.

though the sandstone is of Permian age, around 270 million years old, it is not very tightly cemented. You can rub off sand grains with your hand, and the simple patter of raindrops is enough to erode away a little of the loosened surface sand.

Younger sedimentary rocks have eroded off Cedar Mesa, though some of them are exposed on nearby buttes and mesas: dark red siltstone and mudstone of the Triassic Moenkopi Formation, just above it the white ledge of the Shinarump Conglomerate, and above that the purplish and greenish shales of the Chinle Formation—also Triassic. The mesa to the north is capped with a cliff of red-brown Wingate Sandstone darkened with desert varnish.

The drainage pattern here is controlled by two sets of joints, one trending northwest-southeast, the other northeast-southwest. In many areas within the national monument, horizontal rock surfaces are marked with small potholes that contain little pools of rainwater. As these hollows develop, tiny organisms come to inhabit the pools, and their metabolic processes add acids that further dissolve the rock. As the pools dry up, the organisms curl up in cysts or lay drought-resistant egg clusters that will come to life again when their miniature worlds are once again moistened by rain.

Note the tapered black bands of lichens that mark seepage lines on many cliffs, both on Cedar Mesa Sandstone near the bridges and on nearby cliffs of Wingate Sandstone. Lichens are considered plant-world pioneers in that they initiate processes that break rock down into soil. Shinier blue-black areas are desert varnish, thin deposits of iron and manganese. Smaller

Natural bridges span the streams that give them birth. Here, Sipapu Bridge arches across White Canyon. Halka Chronic photo.

Whirled by streams, sand and larger rock fragments grind out potholes in rocky streambeds. Halka Chronic photo.

patches of desert varnish and more brilliantly colored lichens mark many other rock faces.

Puffy pink soils on the mesa are derived at least in part from the Moenkopi Formation. Gypsum, also from the Moenkopi Formation, is responsible for soil puffiness.

GEOLOGIC HISTORY

Paleozoic Era. The national monument displays no evidence of the earliest history of this area, but most of its Paleozoic story is probably similar to that of the rest of the Colorado Plateau: advances and retreats of a shallow sea across the beveled surface of Precambrian igneous and metamorphic rocks. In Permian time, as the land rose above the sea, coastal sand dunes developed, probably separated (in both space and time) by flat interdune areas where silt and mud accumulated. Their product: the Cedar Mesa Sandstone and its related siltstone and mudstone interbeds.

Mesozoic Era. Early in Triassic time, the sea advanced from the west, depriving the coastal dunes of their source of sand. Here in southeastern Utah, close to the ancient shoreline, on tidal flats and low deltas and in shallow estuaries, the Moenkopi Formation de-

veloped; its mudstone, siltstone, and sandstone layers formed of debris eroded from highlands in Colorado and northern New Mexico. Raindrop marks and mud cracks show that parts of the formation accumulated above sea level.

Later in Triassic time the Chinle Formation accumulated in floodplains, rivers, and shallow, swampy lakes. The climate may have become wetter. Quantities of volcanic ash frequently drifted over the area from volcanoes somewhere to the west, southwest, or northwest. South of here the great logs of the Petrified Forest were buried in muds of the Chinle Formation, and farther east and southeast the oldest dinosaurs left their skeletons in similar muds. Then desert conditions swept the region again, leaving behind the cross-bedded, dune-formed Wingate Sandstone.

No Jurassic or Cretaceous rocks are present here, but we know from surrounding areas that a Cretaceous sea swept across this area, depositing both near-shore and marine sediments.

Cenozoic Era. As the Rocky Mountains rose late in Cretaceous and early in Tertiary time, this area received sediments washed from them. But no Tertiary rocks remain within the monument or even within sight of it.

By Quaternary time, streams draining the west slope of the Rockies wound across a wide, low plain, swinging in easy meanders, stripping away the soft Tertiary sediments. Strengthened by regional uplift within the past several million years, by rapid downcutting as the Colorado River deepened Grand Canyon, and by increased stream flow during the Pleistocene Epoch, streams incised their winding channels, cutting into flat-lying Mesozoic rock layers. As they cut "goosenecks" or entrenched meanders into harder Paleozoic strata—in this region the Cedar Mesa Sandstone—the cliffs of Mesozoic rock retreated, baring the harder Permian unit as the mesa surface. As a final touch, the streams broke through fins of rock enclosed by the goosenecks, creating the natural bridges of this monument.

Prehistoric hunter-gatherers came into this region around 10,000 years ago, but left little sign of their passing. Anasazi peoples, pueblo-builders with a more complex culture, inhabited the area after A.D. 900. Some of them built storage and sleeping rooms in alcoves below the rim of White Canyon, and left rock paintings near the natural bridges.

OTHER READING

Petersen, David, 1990. *Of Wind, Water, and Sand: The Natural Bridges Story.* Canyonlands Natural History Association.

NAVAJO NATIONAL MONUMENT

ESTABLISHED: 1909
SIZE: 0.6 square mile (1.4 square kilometers)
ELEVATION: About 7286 feet (2221 meters) at visitor center
ADDRESS: HC 71, Box 3, Tonalea, Arizona 86044-9704

STAR FEATURES

☼ Prehistoric Indian villages—among the largest in America—protected from the ravages of time by arching overhangs of Navajo Sandstone.

☼ Smaller alcoves and other geologic features characteristic of this formation, visible near the entrance road and on the surface of the Shonto Plateau.

☼ Visitor center, museum, introductory slide show, trails, guided tours to the ruins (summer only). Visitors to the ruins must be accompanied by Park Service personnel. Inscription House is not open to the public.

SETTING THE STAGE

The many-roomed "apartment houses" of Betatakin, Keet Seel, and Inscription House occupy unusually large, deep alcoves in massive, salmon-colored cliffs of Navajo Sandstone. The wide-sweeping cross-bedding and rounded, fine-grained sand of this formation indicate that it is a relic of former sand dunes, part of an ancient Sahara that stretched from southern Nevada into northern Arizona, Utah, Colorado, and Wyoming.

The Navajo Sandstone is one of the major scenery-makers of the Colorado Plateau: It forms the great rock span of Rainbow Bridge, the towering cliffs of Zion Canyon, the bare-rock summits of Capitol Reef, the barren slopes that support the arches of Arches National Park. Here it bends up sharply in the Organ Rock Monocline near the monument entrance road, then levels out into the Shonto Plateau. The formation lends itself well to development of alcoves like those within this national monument, not just the large alcoves of the cliff dwellings but also the many small ones visible along trails to Betatakin and Keet Seel.

Other Jurassic rocks below the massive sandstone—red siltstone and mudstone layers of the Kayenta Formation—also contribute to development of the big alcoves. They are relatively impervious to rainwater and snowmelt that seep downward from the surface through the porous Navajo Sandstone. Along the walls of deep canyons, where the contact between

the two formations is exposed, the water escapes to the surface in lines of springs. And where the Kayenta Formation is kept damp by these springs, it weakens and eventually washes away, undermining the massive sandstone above. Where it is undercut, the Navajo Sandstone breaks away in curving sheets and tumbles into the canyon below. Thus, the alcoves tend to mark the position of springs and seeps—no doubt an important factor in making sites attractive to prehistoric peoples.

Smaller alcoves form in much the same way. Thin horizontal layers of reddish siltstone, thought to represent flat interdune areas where fine dust and silt collected, interrupt the smooth cross-bedding of the Navajo Sandstone. These thin siltstone layers also deflect the downward flow of moisture and initiate development of alcoves. In these small recesses, grains of sand and small rock slabs fall away, commonly leaving cross-bedding surfaces as the alcove ceilings. Animal tracks and droppings show that these alcoves, too, have been used as habitations.

Betatakin Canyon below the visitor center gradually erodes headward by the same process that creates the alcoves: gradual undermining of Navajo Sandstone.

layers along the Sandal and Betatakin Trails are as much as 6 feet (2 meters) thick and contain small nodules of black chert.

In places the sandstone cliffs are streaked with black, mostly due to the growth of lichens along seepage lines, where moisture is a little more plentiful than elsewhere. Don't confuse these dark, V-shaped, nonshiny bands with the shiny, blue-black desert varnish that coats some rock surfaces. Desert varnish accumulates particularly along joints that developed during uplift of Shonto Mesa and of the entire Colorado Plateau region.

Barren slopes near the entrance road and on Shonto Mesa near the visitor center show that many joints, arranged in parallel sets, cut the Navajo Sandstone. One set runs almost north-south, another northeast-southwest. On the mesa surface these joints are deeply incised and serve as watercourses down which tiny rivulets flow during rainstorms. Small pools have developed along them in places, inhabited by tiny organisms that live out their lives in the short days or weeks after a rain, before the pools dry up. Acid by-products from the metabolism of these organisms help to dissolve the calcium carbonate that cements sand grains together, further enlarging the pools. Northeast-southwest joints controlled the position and direction of Betatakin Canyon and several other canyons in this area.

Bare rock slopes are rounded by the gradual peeling off of flakes of rock in a process known as exfoliation. Small patches of red, sandy soil develop in low

Betatakin (below) and Keet Seel (above) occupy large recesses in cliffs of Navajo Sandstone. The sloping ceilings are marked with layering of ancient dunes and streaks of black lichens. Betatakin photo by Halka Chronic. Keet Seel photo by John K. Loleit.

Springs that head the canyon, and the stream issuing from the springs, help to move loosened particles of rock material downstream. The stream may seem too small for such work, but its normal flow is occasionally supplemented by thunderstorm deluges.

Several fine-grained, gray, freshwater limestone layers are present within and at the top of the Navajo Sandstone here. They show us that shallow ponds or lakes developed in some interdune areas. Limestone

Weathering and erosion accent joints that score the surface of the Navajo Sandstone. The roots of pinyon and cactus hold enough soil for an island of vegetation on the otherwise barren rock. Halka Chronic photo.

Tree islands may contain remnants of soil that once covered the plateau surface. Climate changes and grazing (no longer permitted in the national monument) destroyed many of the original plants, whose roots held soil in place. Halka Chronic photo.

spots or at the junctions of joints. Sage, grasses, cactus, pinyon trees, and junipers take possession of these patches; their roots penetrate the joints for nourishment. And while the roots and branches help to conserve the soil, they also break up new soil material.

GEOLOGIC HISTORY

Mesozoic Era. Since none of the early geologic history of this area is evident at Navajo National Monument, we'll start our story in the middle of the Mesozoic Era.

During Jurassic time this part of Arizona was a broad, gentle coastal plain that sloped from the eroded remains of the ancestral Rocky Mountains to the shores of a western sea. The land was low and in Triassic time received fine sediments brought by rivers from the south and southeast. Silt and sand from these rivers would eventually form the red rocks of the Kayenta Formation.

This part of the continent then lay at about the same latitude as the Sahara and other great deserts of today—regions where peculiarities of atmospheric circulation lead to arid climatic conditions. Early Jurassic deserts spread along the coast and far inland, forming a sea of sand 250 miles (400 kilometers) across and 600 miles (1000 kilometers) from north to south. As sand dunes

drifted across the desert region, fine silt and clay accumulated in flat interdune areas. Eventually the sand and interdune deposits became the Navajo Sandstone, about 1000 feet (300 meters) thick here. (See sidebar.)

The region's later Jurassic history was one of changing environments and more flooding by the sea, with stream, lagoon, and marsh deposits as well as marine sediments—a fluctuating sea-coast situation. Some of the sediments carry significant amounts of volcanic ash from volcanoes to the west.

After a long period of erosion, a shallow Cretaceous sea crept in from the east; in it were deposited the rocks that now make up Black Mesa, the tree-covered tableland south of Shonto Mesa. Toward the end of Cretaceous time this sea drained away as the land rose again, signaling the development of the Rocky Mountains.

Cenozoic Era. If Cenozoic rocks were deposited here, they have now eroded away. Drainage in Eocene time was toward large interior lakes in central Utah. With uplift of the Colorado Plateau later in Tertiary time, with its segmentation into smaller individual plateaus, and with development of the Colorado River and the cutting of the Grand Canyon, streams bit ever more deeply into the Plateau surface, some along winding drainage patterns already established. Canyon carving

WHICH SAND IS DUNE SAND?

How do we recognize dune sandstone? Its most obvious characteristic is its broad, sweeping diagonal cross-bedding, formed as wind bounces sand up the dunes' gentle windward slopes and drops it on the steeper leeward slopes. The cross-bedding usually comes in sets of laminae that wedge out at low angles as they intersect other sets. Horizontal stratification is rare, and involves thin brownish or reddish interdune deposits—silt and clay that collect on flats between individual dunes or groups of dunes. The dune sand itself is medium-grained; the grains are all the same size and well rounded. Seen under a hand lens or microscope, individual grains are finely pitted from contact with other grains as they blow in the wind. Dune sand is usually (but not always) composed of quartz grains; if made entirely of quartz, it is white, though often groundwater moving through the sandstone precipitates fine grains of iron oxide, which in varying amounts tint it with shades of pink and red.

Some dune sandstones are quite loosely consolidated, their grains not very well cemented together with minerals brought in by groundwater. Such sandstones usually weather into rounded cliffs, ledges, and domes without sharp corners. The Navajo Sandstone is a good example of this—its rounded surfaces are recognizable all over the Colorado Plateau. When the cementing material is harder—perhaps containing silica—the sand grains fuse together and the sandstones are much more resistant to weathering. With even slight bending such rock breaks along vertical joints, giving us stunning palisades and sharp-edged vertical cliffs like those of the Wingate Sandstone. Flat upper surfaces of both types may be finely decorated with wind-formed ripple marks, and careful searching will sometimes reveal the tracks of long-gone reptiles or crawling insects.

Cross-bedding formed by the action of water is characteristically steeper and straighter, with cross-strata meeting at steeper angles. Trough-style or festoon cross-bedding forms along channels or streams. Wave action in a lake, pond, or sea bevels horizontal erosion surfaces across cross-bedding. In addition, water-deposited sandstone is not as uniform in grain size as sandstone deposited by wind—it may even include pebbles—and the grains aren't as nicely rounded. Of course, on beaches wind and water may join forces, complicating the picture. But if you've spent some childhood summers on sandy beaches, you can recognize the "feel" of a handful of sand eroded from a beach deposit. Much of the Dakota Sandstone fits the bill.

may also have increased in Pleistocene time because of increased precipitation.

Prehistoric people found their way into this area about 10,000 years ago. Hunter-gatherers were followed by Anasazi peoples who built Betatakin, Keet Seel, and Inscription House. These ruins, dated by analysis of tree rings in their roof beams, were constructed around A.D. 1250, then abandoned only fifty years later, probably because of environmental and social problems such as prolonged drought or too frequent raids by their neighbors. The Anasazi may have migrated southeastward to pueblos along the Rio Grande, or southward to the pueblos of the Hopi Mesas. Despite the name of the national monument, they were not related to today's Navajos.

OTHER READING

Viele, Catherine W., 1993. *Voices in the Canyon: The Story of Navajo National Monument.* Southwest Parks and Monuments Association.

PETRIFIED FOREST NATIONAL PARK

ESTABLISHED: 1906 as a national monument, 1962 as a national park
SIZE: 146 square miles (378 square kilometers)
ELEVATION: 5305 to 6235 feet (1617 to 1900 meters)
ADDRESS: PO Box 2217, Petrified Forest National Park, Arizona 86028

STAR FEATURES

- Numerous petrified tree trunks, their wood filled in and replaced, cell by cell, with jasper, agate, amethyst, and other colorful forms of silica.
- Other fossils, including fern, cycad, and ginkgo leaves, reptile (including dinosaur) and amphibian bones, reptile nests, and animal tracks and burrows.
- The Painted Desert, its subtle color patterns a legacy of stream and rivulet erosion in a many-hued palette of siltstone, mudstone, and volcanic ash.
- Visitor centers, explanatory film, museum displays, road guide to points of interest, nature trails, and talks by park naturalists.

See color pages for additional photographs.

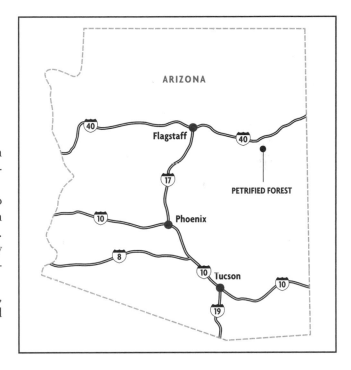

SETTING THE STAGE

The Painted Desert of northern Arizona owes its existence not just to low rainfall, but also to the character of the Triassic rocks that form it. The Chinle Formation, layer upon layer of mudstone, siltstone, and claystone, with liberal additions of volcanic ash, surfaces this desert region. Poorly consolidated, these rocks swell when wet into tacky mud relatively impervious to additional water. Dry, the same surfaces are caked, cracked, and puffy, crunching underfoot and readily crumbling to dust. Unable to anchor themselves in the swelling, shrinking, rapidly eroding soil, few plants survive (see sidebar).

The Chinle Formation, about 220 million years old, is delicately colored with small amounts of iron and manganese compounds and other minerals that occur as minor components of the volcanic clay and mudstone of the Chinle Formation. The soft rocks round into blue, blue-gray, purple, rusty red, white, and yellow mounds and ridges. Where they are capped and protected by hard layers of basalt or by large petrified logs, they stand as flat-topped, steep-sided ridges and small buttes and mesas, also important components of badland scenery. On such steep slopes, particularly just below overhanging ledges of harder rock, the soft layers are better exposed: They are truly rocklike, though they break easily into tiny chips. In some parts of the national park the caprock is lava that came from two low-profile shield volcanoes, Pilot Rock and Pintado Point. Other mesas here, as well as high ones east of the national park, are topped with Pliocene lake deposits.

The Painted Desert badlands extend far beyond the limits of Petrified Forest National Park, stretching southeast toward the White Mountains and northwest along the Little Colorado River almost to Grand Canyon. Where the Chinle Formation has eroded away, deep red siltstone and sandstone of the underlying Moenkopi Formation, whose abundant gypsum also retards plant growth, give them added brilliance.

Petrified logs of this region are exceptionally abundant, exceptionally large, exceptionally colorful, and, unfortunately, exceptionally attractive to rock and mineral collectors. The national park was established with the express purpose of protecting the treasures within it, and collecting any petrified wood—even small fragments—is prohibited. Native Americans also prohibit removal of fossil wood from reservation lands.

RUNAWAY EROSION

In arid and semiarid climates, erosion may create steep gullies, sharp ridges, and rounded, often fancifully colored mounds and hills of badlands. This is especially true of soft, loosely consolidated rocks that are chemically not conducive to plant growth.

Formations that include in their makeup a high proportion of volcanic ash lend themselves particularly well to badland development. Gradual chemical alteration of the volcanic ash produces bentonite, a type of rock including, in particular, a mineral group called montmorillonite, which has the unusual property of expanding to several times its original volume when it gets wet. Then it forms a gooey, gummy, tacky mud that, as it dries, shrinks with a characteristic crisp, popcornlike, exceedingly fragile crust. Expansion and contraction of the montmorillonite prevents plants from becoming established and loosens the surface so that it erodes easily in the next downpour.

Unhindered erosion generates more unhindered erosion. Slopewash on the barren hills sheets into small gullies, and these in turn gather into larger ravines, leaving rounded hills or sharpened ridges between. Erosion of badlands undermines stronger, more resistant overlying strata, which break away in ledges and cliffs, their fragments falling onto the badlands below.

These fragments, as well as nodules, concretions, and petrified bone and wood in the badlands themselves, all cause subtle and not-so-subtle changes in the way the landscape evolves, as do differences in hardness of the strata or the presence of desert pavement—pebbles that concentrate on the surface when finer material washes or blows away. In places, harder, more resistant rock caps spires of softer rock, creating "mushrooms" and "toadstools." A petrified log may ride on a sea of mudstone spires. And throughout many badlands, color banding of different layers adds interest and an otherworldly beauty to the landscape.

Specimens for sale in nearby curio shops come from the same geologic formation outside the park. The colorful badland scenery and petrified logs within the park are photogenic, and photographs make good souvenirs too.

The Petrified Forest is not a forest at all. Most logs are lying on their sides. Many are battered, with limbs and roots broken off and bark scraped away, as if they had been rafted by floods or mudflows, rolled and buffeted, piled into logjams, and then rapidly covered with stream sediments and volcanic ash. Some may have grown nearby; some may have been flood-carried from forested mountains to the south and southwest. Most of the great trunks belong to an extinct conifer species, *Araucarioxylon arizonicum*.

Because volcanic ash is made of tiny fragments of unstable silica glass, groundwater seeping through the sediments soon becomes charged with dissolved silica. The silica tends to come out of solution when it contacts organic material such as old wood or animal bones. Little by little it has accumulated in pore spaces and cells within the trunks, bringing with it traces of iron, manganese, and other mineral substances that now add brilliant color to the wood.

Petrifaction is only one type of fossilization. Fern fronds and leaves from several types of trees occur here as imprints in layers of fine sediment. Tracks and burrows were left by reptiles and other creatures, and include crayfish trails and termite nests. Fossil reptile nests much like those built by today's sea turtles and crocodiles remain in the rock. In some places, snail shells still retain their original calcium carbonate; others have left only molds, the original shell having been dissolved away.

Rafted by flooded streams, battered tree trunks slowly turned to stone as silica impregnated the wood. Silica for fossilization came from volcanic ash, abundant in the badlands behind the log. Ray Strauss photo.

GEOLOGIC HISTORY

Mesozoic Era. The saga of the Petrified Forest and Painted Desert begins in Triassic time, 225 million to 220 million years ago, in marshes and channels of a broad floodplain stretching between high mountains in southern or central Arizona and lower country to the north. Forests of tall conifers bordered shifting waterways and marshes of the floodplain, green with a diverse assortment of mosses and ferns. Grasses and flowering plants had not yet evolved. Living in or close to the water were many types of animals, among them fishes, clams, shrimp, crayfish, crocodile-like phytosaurs, and large awkward-looking amphibians. Many types of insects buzzed and crawled about, nibbling the edges of leaves, forming galls on some of the plants, tunneling plant stems. Other animals roamed between and near the rivers, including mammal-like reptiles, giant amphibians, and dinosaurs. Reptile nests lie along the sides of an ancient river.

At times, floods from the mountains surged across this area, forming new channels, constantly rerouting stream courses, cutting into their banks to topple trees and other vegetation, tumbling the great trunks in their mud-thickened, churning waters. To the west, volcanoes now and then erupted, and cloud after cloud of volcanic ash drifted across the region, adding to the sediments. Floodwaters carried fallen trees and other vegetation, buried their battered trunks (many with root stumps still attached), and swept over them layers of ash and coarse sand and pebbles washed from the mountains. From time to time, standing trees of the floodplain were inundated and preserved where they stood. More layers of silt, sand, and clay, and more blankets of ash from more eruptions, accumulated above the entombed trunks.

The trees were buried so quickly that they did not decay. Instead, as silica-laden waters flowed through them, their pore spaces were little by little filled with silica. Where iron or manganese were present, some of the petrified wood took on subtle hues of agate and brilliant tones of jasper—white and yellow, red and black. In hollows within the trunks, quartz crystals had room to grow, some glassy and colorless, others tinted the pale lavender of amethyst.

Ultimately the mountains to the south wore away. In Cretaceous time, seas advanced again, coming not from the west as in Paleozoic time, but from the east. New deposits were laid down, thick shale and sandstone layers bearing fossils of marine creatures rather than of land animals.

About 70 million years ago, as the Cretaceous Period and the Mesozoic Era drew to their close, this region began to rise once more and the seas drained away for the last time. As the earth moved, the stresses and strains of uplift cracked many of the buried, petrified trunks and segmented them like cordwood.

Cenozoic Era. For much of the time since the close of the Cretaceous Period, erosion has been the dominant force here. Though the region has been at times grassland, at times marshy floodplain, at times covered with lava and new falls of volcanic ash, its Cretaceous marine sediments were gradually washed

Scattered logs, naturally broken into cordwood lengths, lie in deeply eroded ravines completely lacking in plant growth. Ray Strauss photos.

away. Then for millions of years the region held a great lake or a number of small lakes, in an area as large as Lake Erie today. In the waters, lake sediments accumulated, forming the Bidahochi Formation, now visible at the park's northeast boundary. Volcanoes were active nearby, contributing lava and volcanic ash, some of them erupting with violent bursts of steam through the muddy bottom of the lake. Then erosion gained the upper hand again, removing much of the lake sediment and lava and volcanic ash, and baring at last the old Triassic floodplain deposits: sandstone and mudstone and conglomerate, colorful bentonite, fossil reptiles and amphibians, insect impressions and leafprints, and the petrified logs of this national park.

BEHIND THE SCENES
Viewpoints and loops are described in north-to-south order, as they appear on entering the park from the north.

Painted Desert Loop. Observant eyes will see many geologic features here: badlands eroded in soft, bentonite-rich rock layers; subtle colors given to the Painted Desert by the minerals of nature's palette; ridges and mesas protected by hard sandstone layers and thin, relatively recent lava flows; low-profile volcanoes from which some of these lava flows came; undermined lava blocks breaking away from the mesas; sharp white veins of gypsum; deposits of white caliche.

Some of the young volcanic rocks along the mesa edge and near the picnic area are simple lava flows with many gas bubble holes or vesicles. Others include large blobs of basalt thrown forcibly from volcanic vents, rounded and football-shaped from their brief flight through the air.

Geologists believe that the gypsum layers formed here in shallow, evaporating lakes and ponds of the Triassic floodplain. Gypsum is a common constituent of many rock layers in the Southwest and dissolves fairly easily, so the streams feeding these ponds may have contained large amounts of the mineral. Gypsum in veins is a secondary deposit, leached by groundwater from gypsum layers within the rock and redeposited in slender cracks. The gypsum occurs as a transparent, silky-surfaced mineral called selenite. Salt, which also may have been deposited in the evaporating lakes, dissolves readily; if it formed here in Triassic time it has been removed by groundwater.

Newspaper Rock. This site holds interest for geologists as well as archaeologists because the petroglyphs were created by patiently pecking through the dark desert varnish of the rock face. Desert varnish is a shiny brown or blue-black coating of manganese and iron minerals derived from airborne dust or from the rock itself. Rain or other moisture dissolves and carries along tiny amounts of these minerals. Later, as the moisture evaporates from the sun-warmed rock, it leaves its mineral burden, however small, at the surface. As you can see here, where petroglyphs were pecked through it hundreds of years ago, new desert varnish has not yet had time to form.

Blue Mesa Loop. The approach from the main park road to Blue Mesa heads almost straight toward a small, low volcanic cone, a shield volcano that emitted very fluid basalt lava—hence the low profile of the cone.

A layer of cross-bedded sandstone, one of the coarser, more resistant layers of the Chinle Formation, caps Blue Mesa. Short, straight cross-bedding in this layer is of the type formed by running water, so we know that the sandstone was deposited in a stream channel or along the edge of a small delta, with layers of sand building downstream.

East and south of the south parking area, in a maze of little canyons, petrified logs perch on narrow ridges, for a time sheltering the soft rock that makes up their pedestals. When the pedestals finally wash out from

Painted hills of Chinle Formation, rich in clays formed from volcanic ash, are too unstable for plant growth. The soft clays and fossil "forests" are partly the result of explosive volcanic eruptions. National Park Service photo.

under them, the logs will roll or tumble downslope. This "forest," with its large number of tree trunks, is thought to have been a logjam in the channel of a flooding Triassic stream. The blue, white, lavender, and pale green of eroded slopes remind us that this region is part of the Painted Desert. These colors form where oxygen is in short supply in the original sediment, as it often is in ash falls or in floodwaters containing abundant decaying plant and animal matter.

The view north from Blue Mesa reveals a different scene, with tones of brown and rust predominating, and with the edge of the mesa reaching north like the paws of a giant sphinx. Single layers in the Chinle Formation are not extensive and are in many places replaced horizontally by strata of different color, grain size, or bentonite content.

The loop trail follows the summit of a ridge and then drops downward close to these rocks. Notice the desert pavement along the ridge, with small round pebbles of hard igneous and metamorphic rock or gray chert. The pebbles were washed here in Triassic time from highlands in central Arizona, about 60 miles (100 kilometers) south and southwest of here. The chert comes from Permian rock—the Kaibab Formation, now absent in central Arizona. Igneous and metamorphic pebbles are mostly Precambrian, and provide important evidence of the existence of high mountains there in Triassic time. Studies show that the pebbles become larger southward toward their source.

From the trail, watch for large slump blocks that are sliding down and away from the edge of the mesa. Some tilt back toward the mesa, showing that they moved on curving slide planes.

Agate Bridge. At this site a petrified log spans a small, usually dry watercourse. The ends of the log rest in the sandstone in which it was originally entombed. The concrete support was added in 1917, when it was feared the heavy stone log might collapse. Several other log spans occur in the park.

Jasper Forest. A thin layer of pebbles covers much of this area. As at Blue Mesa, many of the pebbles are of recognizable Precambrian metamorphic and igneous rocks, or of chert from the Kaibab Limestone. Washed in Triassic time from mountains in central Arizona and deposited as conglomerate in the Chinle

Formation, they have now been recycled into desert pavement. Other pebbles are small angular fragments of petrified wood. The pebbly pavement is in places cemented by hard white caliche, the "hardpan" of desert regions, deposited near the surface as lime-laden groundwater evaporated. Cemented or not, the desert pavement protects underlying clay and siltstone from erosion. It also helps rainwater sink into the ground rather than flow off across the surface, and forms an armor that lessens evaporation.

Crystal Forest. The name of this "forest" stems from crystals of quartz and amethyst in cavities of the petrified logs. Unfortunately, most of these crystals were extracted before the area came under national park protection. Mineral collectors went so far as to blast open many trunks to get at the clear, well-formed crystals.

Flattops. The pebbly surface here does not everywhere coincide with a hard caprock. In Pliocene time a stream's broad floodplain covered this part of the national park; the stream and its tributaries later cut down into this surface, leaving only the present flat-topped mesa as a remnant of the old floodplain. The increased erosion may have been due to rapid downcutting in Grand Canyon and in the gorge of the Little Colorado River, as well as to increased precipitation during Pleistocene rainy cycles.

Rainbow Forest and Museum. Exhibits here display fossil leaves and reptile bones found in the park, as well as polished petrified wood showing its vivid colors and the microscopic details of the original wood.

The museum is near an unusually dense concentration of large petrified logs, many of them brilliantly colored with red and yellow jasper. Some show remnants of their original bark—often a clue, as the exhibits show, to the kinds of trees involved.

OTHER READING

Bezy, John V., and Arthur S. Trevena, 2000. *Guide to Geologic Features at Petrified Forest National Park*. Arizona Geological Survey.

Colbert, E. H., and R. R. Johnson, 1985. *The Petrified Forest through the Ages*. Bulletin series 54, Museum of Northern Arizona Press.

Long, Robert A., and Rose Houk, 2000. *Dawn of the Dinosaurs: The Triassic in Petrified Forest*. Petrified Forest Museum Association.

RAINBOW BRIDGE NATIONAL MONUMENT

ESTABLISHED: 1910
SIZE: 0.25 square mile (0.64 square kilometer)
ELEVATION: Lake Powell high-water level, 3700 feet (1128 meters)
ADDRESS: c/o Glen Canyon National Recreation Area, PO Box 1507, Page, Arizona 86040-1507

STAR FEATURES

- ✪ The world's largest known natural bridge, carved by stream erosion amid a landscape of barren cliffs and canyons.
- ✪ Sedimentary and erosional characteristics of the dune-formed Navajo Sandstone, as well as those of rock formations above and below it.
- ✪ Views of Navajo Mountain, a large laccolith.
- ✪ Visitor center at Glen Canyon Dam with exhibits and introductory film; boat dock and short trail to Rainbow Bridge.

SETTING THE STAGE

More accessible now than it was before Lake Powell filled, Rainbow Bridge, deep in a slickrock wilderness of rugged canyons, arches across the channel of Bridge

Streaked and stained with desert varnish, the sculptured arch of Rainbow Bridge, carved in Navajo Sandstone, rests on narrow ledges of resistant sandstone. Halka Chronic photo.

Creek, one of several streams that drain the north slope of Navajo Mountain. The bridge, whose graceful 275-foot (84-meter) span soars 290 feet (88 meters)—the height of a 30-story building—above the creek, is the largest rock span known.

Formed of salmon-colored Navajo Sandstone, a rock unit responsible for much of the scenic grandeur of the Plateau country, the bridge owes its shape to sedimentary and erosional characteristics of this sandstone. Long, sweeping cross-bedding, coupled with uniformly fine, rounded, frosted sand grains, shows that the formation is wind-deposited—the product of desert sand dunes. Sand grains are cemented together with calcium carbonate to form a strong sedimentary layer capable of standing in high vertical cliffs. En route to Rainbow Bridge by boat or by trail it is easy to see such cliffs, marked here and there with large, arching overhangs.

Above the Navajo Sandstone are thinly layered shale and siltstone of the Carmel Formation. This unit, best seen from Lake Powell and in road cuts between Glen Canyon Dam and Wahweap, is weak and easily eroded, wearing back into broad benches.

Above it, the Entrada Sandstone caps many of the buttes and promontories that jut into Lake Powell. Like the Navajo Sandstone, the Entrada is an eolian (wind-formed) rock unit, complete with large-scale diagonal cross-bedding and fine, well-rounded sand grains.

Below Rainbow Bridge, and contributing to its genesis, is the Kayenta Formation, a series of red sandstone and siltstone ledges that bear the features of river floodplain or delta deposits. In many places this rock is covered by lake waters; at Rainbow Bridge it is above water level and forms the horizontal ledges that support the bridge. Since its sandstones are hard and resistant, the Kayenta Formation as a whole erodes less easily than the overlying Navajo Sandstone. It is also less porous than the Navajo Sandstone, and deflects rainwater and snowmelt percolating down through the sandstone, causing springs and seeps along the contact between the two formations.

South of Rainbow Bridge, visible up Bridge Canyon, Navajo Mountain rises to 10,388 feet (3166 meters) above sea level. It, too, probably played a role in creation of the natural bridge. Navajo Mountain is a laccolith, a granite intrusion that about 65 million years ago pushed between layers of sedimentary rock, never reaching the surface but doming the rock units that lay above it. Strata that once covered Navajo Mountain like draped blankets have eroded from its summit, but their upturned edges still encircle its base.

How and why did Rainbow Bridge form? A single, often overlooked feature near the bridge gives us a

Thin-bedded, slabby, and resistant, the Kayenta Formation forms a precarious base for Rainbow Bridge. Ray Strauss photo.

good clue. Behind the east buttress of the bridge, well above the present stream channel, part of an abandoned older channel tells us that Bridge Creek once took a sharp loop here, a "gooseneck" meander inherited from a slow, sluggish course across a gently

A line of seeps, where mosses and lichens add black accents to the tawny rock, weakens a Navajo Sandstone cliff. A curving fracture (arrows) is developing above this weakened zone. The rock below it will eventually drop away, leaving an arching recess. Halka Chronic photo.

Bridge Creek under Rainbow Bridge shows clearly that this is a bridge rather than an arch. Photo courtesy of Northern Arizona University, Cline Collection.

sloping surface. As the stream—invigorated by plateau uplift, downstream downcutting, and the rising mass of Navajo Mountain—cut ever deeper, it retained its gooseneck pattern, entrenching its meander loop deeply into the rock.

As it rushed through this twisting gorge, the stream pounded sand, pebbles, and boulders against the rock spur circumscribed by the entrenched loop. At some stage, as the spur became thinner and thinner, the stream broke through it, abandoned its entrenched gooseneck, and took the more direct route downstream.

At first the bridge was probably thick-bodied and heavy, with only a small opening. But, helped by its steepened gradient, the stream continued to cut downward through the Navajo Sandstone. The resistant Kayenta Formation, below the Navajo, retarded downward cutting for a time, and the stream instead widened the passage below the bridge. Springs along the contact between the formations probably aided in the erosional process by keeping the rocks well saturated and subject to freeze-and-thaw weathering; they may also have helped to remove some of the cementing material that held the sandstone together.

Perhaps it was at this stage that large blocks of Navajo Sandstone fell away in the archlike pattern common in this area. With its opening thus enlarged, and with further deepening of the stream channel, bridge surfaces were further refined. Thin, curving sheets of rock spalled off them, gradually smoothing the great span—a process that will continue until the bridge becomes so narrow it can no longer support its own weight.

GEOLOGIC HISTORY

Since rocks of the Precambrian and Paleozoic Eras, deeply buried here, play little or no part in the origin of Rainbow Bridge, we will start our story in Mesozoic time.

Mesozoic Era. Three rock units are exposed at or near this national monument: the Kayenta Formation, the Navajo Sandstone, and the Carmel Formation, the last high up, just out of sight from the canyon floor, and not within monument boundaries. These units tell us of the changes from the floodplain or delta environment of the Kayenta to arid desert, where the light-colored, cross-bedded Navajo Sandstone, prone to erode into domes and arches, was deposited in a dune field rivaling today's Sahara. Following their formation, thousands of feet of Jurassic and Cretaceous sedimentary deposits accumulated, some on land, some during pulselike advances of an eastern sea.

Cenozoic Era. More thousands of feet of sedimentary rocks may have accumulated here in Tertiary time: thick sandstones and siltstones made of rock material brought by streams draining the newborn Rocky Mountains. As their gradients steepened, tributary streams that had meandered sluggishly across a low-elevation plain became vigorous enough to strip away whatever younger sediments may have existed, and to incise their earlier gooseneck meanders into the underlying Mesozoic rocks. Stream vigor in the Rainbow Bridge area was further enhanced by upward doming of the Navajo Mountain laccolith about 65 million years ago, and by later downward erosion by the Colorado River.

As this great river and its small tributaries cut deep into Jurassic and then Triassic rocks, many narrow, winding canyons formed, including mystically beautiful Glen Canyon on the Colorado River, now hidden beneath Lake Powell's waters. Twisting salmon-colored cliffs and slopes of barren rock, etched by wind and rain, displayed the diagonal cross-bedding of Jurassic dunes and the brown and purple streaks of desert varnish. After the rainy cycles of Pleistocene time, this area once more became a desert, with sand from deserts of the past becoming ornaments in a desert of the present.

OTHER READING

Ladd, Gary, 1998. *Rainbow Bridge: The Story Behind the Scenery*. KC Publications.

SUNSET CRATER NATIONAL MONUMENT

ESTABLISHED: 1930
SIZE: 4.8 square miles (12 square kilometers)
ELEVATION: 6958 feet (2120 meters) at visitor center
ADDRESS: 6400 N. Hwy 89, Flagstaff, Arizona 86004

STAR FEATURES

✪ A classic cinder cone, geologically young and of near-perfect symmetry, with smooth gray slopes tinted at the rim in sunset colors. Jagged lava flows nearby are still fresh-looking and almost unvegetated.

✪ Spatter cones, squeeze-ups, lava bubbles and tunnels, and other detailed volcanic features.

✪ Views of other cones of the San Francisco volcanic field, with San Francisco Mountain as the reigning monarch. All the features of this field developed during the last 6 million years.

✪ Interpretive film and exhibits, self-guided nature walks, evening programs.

SETTING THE STAGE

Sunset Crater greets visitors with a tinted summit due not to evening light (though a colorful sunset heightens the effect) but to oxidation and fumarole (steam vent) deposits at its rim. The cone of dark gray basalt cinders rises 1000 feet (300 meters) above its surroundings. Except for a lopsided rim caused by the leeward drift of cinders and ash before prevailing southwest winds, it is symmetrical in shape. A blanket of cinders and ash jettisoned by Sunset Crater extends far east, north, and south, covering an area of almost 800 square miles (2000 square kilometers).

The basalt cinders, small pebbles of bubble-filled lava, have not weathered significantly in this cool, dry climate; they are still crisp and crunchy underfoot. Nor is the crater's unprotected surface grooved by gullies, as you might expect, for it is so porous that rain and melting snow sink right in.

Two rough, black lava flows lie at the foot of the crater. The western or Bonito flow, the one most visitors see, fills the basin between older cinder cones. The Kana-a flow on the east side of the crater is long and narrow because it flowed down a preexisting stream channel. Where it is exposed to view, the lava looks as though it oozed only yesterday from fissures at the base of the cone. But look again: The lava surfaces are spotted with lichens, pioneers of the plant world. With their

help, and with increasing assistance from other plants, the rock will eventually change to soil. Shrubs and trees are already taking root in pockets of cinders and wind-blown pine needles. Each new tree or shrub or clump of grass will add to the growing supply of humus when it drops its leaves or needles, or dries in autumn.

In places the lava contains pieces of white or light gray Paleozoic limestone broken from the walls of the volcano's conduit as magma moved upward. Blocks of intrusive rock have been brought up, too, from intruded masses related to much older, subsurface igneous activity.

Flow patterns are well developed on the Bonito lava flow. Look for ropy **pahoehoe** lava; for rough and hard-to-walk-on **aa** lava, for patterns of concentric pressure ridges curving across lava tongues. Both flows and cinders are basalt and came from the Earth's mantle, rising along a fissure. Because it cooled fairly rapidly, the rock is quite fine grained, except for scattered little crystals of green olivine and white or transparent feldspar.

On the lava flows, squeeze-ups are a late development. Their parallel grooves and scratches record the manner in which they pushed up through jagged breaks

A classic cinder cone, Sunset Crater erupted around A.D. *1100. Its colorful rim hides a deep central crater. Tad Nichols photo.*

in the hardening skin of the lava flows. Lava caves and tunnels developed where lava flowed out from beneath its own hardening crust. Blisters and bubbles on the lava surface, some of them several meters across, formed as hot gases pushed upward against a still flexible skin of cooling lava.

Trending northwest and southeast from Sunset Crater, a line of spatter cones defines the fissure from which the main crater's cinders and lava were emitted. Red-tinted spatter from these cones clearly overlies the black cinders of the parent crater, so we know that they represent a fairly late stage in the eruption story.

O'Leary Peak, north of Sunset Crater, is another expression of volcanism, a lava dome squeezed upward about 200,000 years ago as a doughlike mass of very thick, viscous lava. Considerably lighter in color than Sunset's dark flows, its lava is more silicic than that of Sunset Crater; in composition it resembles lavas of the San Francisco Peaks. The view from the summit takes in Sunset Crater, the Bonito flow, many other cinder cones, and the eroded stratovolcano of San Francisco Mountain itself. Eastward, O'Leary Peak overlooks the eastern edge of the Coconino Plateau, where Paleozoic sedimentary rocks bow downward and plunge beneath younger strata of the Painted Desert. (Wupatki National Monument lies on this great flexure; the story of its prehistoric villages ties in with the story of Sunset Crater's eruption.)

As volcanic pressures build up under a lava crust, pasty lava pushes through cracks to form squeeze-ups. Tad Nichols photo.

GEOLOGIC HISTORY

Precambrian Time, Paleozoic and Mesozoic Eras. The Precambrian and Paleozoic history of this region is essentially identical to that of Grand Canyon, known from unequaled exposures there. Here in the San Francisco volcanic field, these older rocks are concealed by a veneer of lava flows and volcanic cinders. Dark red Triassic sandstone and mudstone originally covered this area too, as did several thousand feet of Jurassic and Cretaceous sandstone and shale, relics of ancient deserts and floodplains and one last incursion of the sea.

Cenozoic Era. With uplift of the Colorado Plateau in early Tertiary time, erosion became the main process shaping most of the landscape. Particularly attacking the Kaibab Uplift, one of the highest plateau segments, Cenozoic erosion stripped off layer after layer of Mesozoic sandstone and shale, laying bare the hard Paleozoic limestone below—a process still going on in the Painted Desert to the east and north.

But about 6 million years ago, during the Pliocene Epoch, volcanic rumblings warned that other landscape-formers were gathering strength far below the plateau surface. The first volcanic outpourings were of very fluid basalt that spread like a thin veneer over the Coconino Plateau, the southern part of the Kaibab Uplift. Then, about 1.8 million years ago, there followed a series of eruptions of thicker, more silicic lava, fairly light in color because of its high quartz and feldspar content. Lava flows and volcanic ash from innumerable eruptions piled up gradually into the stratovolcano of San Francisco Mountain. Lava domes and in some places laccoliths pushed upward, too. Finally, late in the history of the volcanic field, long after the San Francisco volcano had been reshaped by explosion and Ice Age glaciers, some 400 cinder cones erupted around its base.

Sunset Crater is the youngest of these cinder cones. Its cindery slopes and jet black flows are less than a thousand years old. We can readily reconstruct its birth and growth because Sunset Crater closely resembles a modern volcano, Paricutin, which erupted in west-central Mexico between 1943 and 1954. Paricutin was born before the astonished eyes of a Mexican farmer as a tiny fissure in his cornfield began to emit strange

San Francisco Mountain's Inner Basin, roughed out by explosion or collapse, was smoothed by Pleistocene glacial erosion. Sunset Crater cinders blanket the foreground. Tad Nichols photo.

rumbles and belch thin wisps of smoke. Becoming bolder, the new vent tossed out a few bits of spongy pumice—then more, and more, until a cone began to form. Soon cinder and ash and spinning football-sized bombs of lava or lava-coated rock rained down mercilessly on cornfield, house, and surrounding country, killing all vegetation and driving local inhabitants from their homes. The new volcano of course attracted geologists and photographers who recorded its growth in lasting detail. As frothy magma rose and was blown from the summit in a long succession of explosions, less gaseous magma began to rise in the conduit and eventually broke out through the base of the cone and streamed over the ash-covered countryside, engulfing finally the entire village of Paricutin.

Sunset Crater, too, was witnessed by humans (see sidebar). It, too, deluged cornfields as well as plots of beans and squash, staple crops of early Sinagua people who made their homes in this region. Several prehistoric pit houses have been unearthed by archaeologists, their roofs collapsed by the weight of the cinders. Tree-ring studies of charred roof timbers from these dwellings, as well as radiometric and paleomagnetic dating of the volcanic rock, tell us the cinder cone first erupted sometime around A.D. 1100 and was

active on and off for the next 200 years. Like Paricutin, Sunset Crater lies in a setting studded with older cinder cones. As at Paricutin, lava flows oozed from its base. Here the Kana-a flow, northeast of the crater, developed at about the same time as the first main cinder eruption. The Bonito flow came later, and is variously dated as A.D. 1180 or 1220. Smaller eruptions later flung cinders onto the surfaces of these flows. Spatter cones formed as the cinder activity subsided. Near Sunset's crest, volcanic gases escaped from fumaroles for some time after other activity ceased, perhaps until 1220 or 1250, depositing gypsum and sulfur and reddening the rim with iron oxides, the last gaseous gasps of the dying volcano.

There is no reason to believe that the eruption of Sunset Crater brings volcanism to a close here. Dating of other cinder cones in the San Francisco volcanic field shows that they become progressively younger toward the northeast, as if the Earth's crust had drifted slowly southwest over a stationary "hot spot" in the mantle. So another eruption may take place, probably in the area northeast of Sunset Crater. When it comes, it will give warning in the form of earthquakes, steam vents, and perhaps swelling and cracking of the surface.

Spatter cones build up as fluid lava splashes to the surface. Tad Nichols photo.

DISPLACED PERSONS AND A POPULATION EXPLOSION

Sunset Crater's story involves more than lava flows and a blowout of volcanic cinders. Archaeologists working on sites within and near the San Francisco volcanic field, including Sunset Crater and Wupatki, have realized for many years that the eruptions of the crater were witnessed by Sinagua peoples of this region, who must have reacted much as did the farmer witnessing the birth of Paricutin. Local inhabitants must have backed away, startled and probably frightened. Soon, as their fields were devastated and their simple dwellings destroyed, they fled the area and searched for new homes beyond the volcano's reach. Years later—perhaps many generations later—when eruptions of both lava and cinders quieted, they or their descendants returned to find that crops grew well in the cooled cinders and ash. Perhaps the ash had a mulching effect that curbed evaporation and conserved precious moisture, thereby adding to the productivity of fields and garden plots. In an early population explosion, they settled here in ever-greater numbers, building clustered, multistoried apartment houses among the rocky fringes of the cinder fields, houses such as those at Wupatki National Monument. A cooler, moister climate, indicated by growth rings of trees, may have helped to open up this area to increased farming.

Until a few years ago, this was the picture drawn by archaeologists. Recently, scientists have deduced that much broader changes took place among the returning Sinagua—changes in the way villages were set up, increasing sophistication in building construction, changes in trade routes, and evidence of intermingling of different cultural groups— all pointing to the Sunset-Wupatki area's increasing importance as a cultural and perhaps religious center.

Was this importance due in part to the eruption of Sunset Crater? Was the crater significant for cultural or religious reasons in addition to the mulching effect of its black cinders? Archaeologists now suggest that it was, and that evidence of corn offerings to the cinder cone, patterns of population change, and tales passed down through Hopi and other modern groups all indicate its increased importance to inhabitants of the region.

OTHER READING

Houk, Rose, 1995. *Sunset Crater National Monument.* Southwest National Parks and Monuments Association.

Moore, R. B., and E. W. Wolfe, 1987. *Geologic Map of the East Part of the San Francisco Volcanic Field, North-Central Arizona.* USGS Miscellaneous Geologic Investigations Map MF-1960.

Thybony, Scott, 1987. *A Guide to Sunset Crater and Wupatki.* Southwest National Parks and Monuments Association.

VERMILION CLIFFS NATIONAL MONUMENT

ESTABLISHED: 2000
SIZE: 458 square miles (1187 square kilometers)
ELEVATION: 3100 to 6500 feet (945 to 2000 meters)
ADDRESS: BLM Arizona Strip Field Office, 345 E. Riverside Drive, St. George, Utah 84790-9000

STAR FEATURES

✪ Thirty miles of a dramatic escarpment of Triassic and Jurassic strata, rising nearly 3000 feet (nearly 1000 meters) above Paleozoic rocks of the Marble Platform.

✪ The Paria Plateau, with elevations above 7000 feet (2000 meters).

✪ The Paria River, a Colorado River tributary, whose deep, rock-walled canyon lures hikers from all over the world.

✪ Extensive exposures of Mesozoic rock, including unusual erosional features such as the tepee-shaped spires of Coyote Buttes on the Paria Plateau.

✪ The Paria Contact Station, offering maps, brochures, and backcountry permits and advice. Trails and unimproved roads access remote parts of the monument.

SETTING THE STAGE

The high escarpment of Vermilion Cliffs dominates the scenery along US Highway 89A from Lees Ferry and Navajo Bridge to House Rock Valley, where the cliffs turn northwestward and then northward toward the Utah line. The cliffs, 2000 to 3000 feet (600 to 900 meters) high, are made up of Triassic and Jurassic sedimentary rocks varying in color from dark red to salmon pink. Their differing resistance to erosion gives the cliffs their shape: More resistant sandstones break away into vertical walls; less resistant interbedded sandstone, siltstone, and mudstone develop into steep, ledgy slopes; soft, easily eroded limy mudstone, mixed with volcanic ash, forms low rounded hills.

The cliffs and the Paria Plateau are topped with Navajo Sandstone, the white-to-salmon-colored sand dune deposit whose sweeping cross-beds and round-edged pattern of weathering shapes much of the Plateau's striking beauty. The Kayenta and Moenave Formations, and in places the Wingate Sandstone, make up the red ledges and cliffs below the Navajo Sandstone. Together, these four formations make up the Glen Canyon Group.

Colorfully banded, purplish hills along the southern and southwestern edges of the cliffs are the Triassic Chinle Formation, whose liberal amounts of volcanic ash, converted into a group of minerals known as bentonite, tend to swell when wet and shrink when dry, greatly inhibiting plant growth (see sidebar on page 132). Further erosion of the Chinle Formation by both water and wind leads to development of smoothly rounded badland hills. A hard, ledge-forming conglomerate at the base of the Chinle Formation separates it from another Triassic unit, the Moenkopi Formation, which borders US Highway 89A for some distance and lies directly on the Kaibab Limestone of the Marble Platform.

These are erosional cliffs, worn back from the canyon cut by the Colorado River, and erosion is still at work here. As less resistant rocks beneath them erode away, cliff-forming sandstones are undermined. Rockfalls are common, adding fallen fragments to the slopes and building talus cones along the base of the cliffs. In places great masses of layers slid downward along curving slide planes at the front of the cliffs, so that their once horizontal bands now tilt back toward the cliffs. In winter, water freezes in joints in the

144

Shinarump Conglomerate caps the ridge in the foreground; on the main cliff behind, light-colored Navajo Sandstone tops darker Moenave and Kayenta Formations. Lucy Chronic photo.

sandstone and, expanding as it freezes, helps the process along. Here and there, large blocks of fallen sandstone or conglomerate protect softer rock directly beneath them, developing into top-heavy mushroom rocks. Gradually, almost imperceptibly, the cliffs retreat northward and northeastward.

At their eastern end, near Lees Ferry, the Vermilion Cliffs are cut off by the narrow canyon of the Paria River, a Colorado River tributary. Famous as a hiking route (safe only in dry weather), this canyon has tributaries of its own, many of them narrow "slot canyons" walled by smooth-surfaced waves of pink and white sandstone. During the summer rainy season, thunderstorms sweep above the slot canyons and send turbulent, sand-laden waters that scour the smooth rock walls and carry away loosened rock material, bits of vegetation, and anything else in their paths.

Near its outlet, the Paria River flows across the soft Chinle Formation, here bent down to the east by the Echo Cliffs Monocline, one of several such monoclines on the Plateau (see sidebar on page 112). For some distance northward, the monocline parallels the national monument boundary.

GEOLOGIC HISTORY

Paleozoic Era. No Paleozoic rocks are exposed in the national monument, though they underlie all of the Paria Plateau. Permian limestone of the Kaibab Formation is visible on the Marble Platform just south of the Vermilion Cliffs, and to the west along the East Kaibab Monocline, where it bends downward from the surface of the Kaibab Plateau. Its buff-colored ledges can also be seen in Marble Canyon at Lees Ferry or below Navajo Bridge. In this area it is siltier and sandier than it is on the Kaibab Plateau or at Grand Canyon, where it was deposited farther from the Permian shoreline.

Mesozoic Era. Triassic and Jurassic rocks in this region were deposited on a broad plain that sloped northwestward toward the sea. The deep red-brown sandstones and mudstones of the Moenkopi Formation, forming ledges and shaly slopes, hint at their own origin—sand and mud eroded by streams and rivers from the Precambrian cores of the Uncompahgre and Defiance Uplifts in Colorado and New Mexico, and other uplifts farther away. These rivers carried with them iron-bearing minerals—notably biotite and hornblende—from ancient granite, gneiss, and schist. The iron was redeposited as minute particles of hematite intermingled with grains of quartz sand and silt, also derived from the Precambrian rocks of the mountains. Though mountain-derived, the Moenkopi here was deposited close to sea level; its near-shore deposits intertongue with marine limestone not far west of here.

The Moenkopi also bears many indications of the

A boulder fallen from the resistant ledge of Shinarump Conglomerate protects a mudstone pedestal.
Lucy Chronic photo.

environment in which it was deposited, suggesting a warm, arid or semiarid climate. Current ripples, wave ripples, and shrinkage cracks are fairly common; less abundant are arthropod and amphibian tracks, raindrop impressions, and casts of salt crystals. Gypsum present in parts of the formation developed as salty ponds and lagoons evaporated.

The Moenkopi Formation is topped by a prominent dark brown ledge of coarse, pebbly conglomerate, the Shinarump Conglomerate, considered to be the lowest member of the Chinle Formation. The conglomerate extends for many miles south and west of here, and may have accumulated in a broad "paleovalley" on the Triassic plain, where braided, often flooded streams spread its stream-rounded pebbles and sand.

The rest of the Chinle Formation appears to have been deposited on a low, almost featureless plain by sluggish, meandering streams flowing northward toward marshes and perhaps lakes in Utah. Its color suggests a low-oxygen, marshy environment with decaying plant material. Volcanic ash, presumably from volcanoes to the south and west, contributes to its lack of vegetation

and its swelling-when-wet tendency, described above. The colorful layers extend south beyond Petrified Forest National Park.

Above the Chinle, the Moenave and Kayenta Formations, the lower part of the Vermilion Cliffs, are also primarily stream and river deposits. Upward, they intertongue with and give way to thick dune deposits of Navajo Sandstone, demonstrating the sporadic advance of a sea of sand dunes across the nearly flat surface. The Navajo tops most of the Paria Plateau, where some of it is gradually being reworked into modern dunes.

As on most of the Plateau, a thick blanket of additional sediments once covered this area. With Tertiary uplift, streams and rivers gradually washed away Cretaceous and later deposits and cut deep canyons in older rocks. Recent deposits—the sand of canyon floors, rockfall and landslide debris below the cliffs themselves, the debris filling House Rock Valley—give evidence of continuing erosion.

OTHER READING

Kelsey, Michael R. 1998. *Hiking and Exploring the Paria River.* Kelsey Publishing.

WALNUT CANYON NATIONAL MONUMENT

ESTABLISHED: 1915
SIZE: 5.6 square miles (14.5 square kilometers)
ELEVATION: 6800 feet (2072 meters) at visitor center
ADDRESS: 6400 N. Hwy 89, Flagstaff, Arizona 86004

STAR FEATURES

- ✪ Small cliff dwellings in recesses in Permian limestone layers similar to those in the upper walls of Grand Canyon.
- ✪ A deeply entrenched loop of Walnut Creek.
- ✪ A close look at the near-shore version of two formations and at weathering features in limestone and dolomite.
- ✪ Visitor center, slide show, illustrated trail guides, guided walks.

SETTING THE STAGE

One of many small gorges that lace the Coconino Plateau, Walnut Canyon provides an interesting view of geologic processes of Permian time, as well as of the more recent past. Around 800 years ago, at the time when Sinagua Indians built homes in cliff recesses in the canyon, Walnut Creek flowed more often than it does now, and may have retained water in perennial pools even when the stream was dry. Now most of its flow is impounded in Lake Mary as part of Flagstaff's water supply.

Immediately below the visitor center the stream describes a tight loop that may originally have been a meander bend across a relatively even, low-gradient surface, or that may have originated in a twisting solution cavern in the limestone, a cavern that later collapsed. In either case, as its gradient steepened the stream retained its looplike course but cut deeper and deeper—more than 350 feet (100 meters) in all—into surrounding rock layers.

From the visitor center the view looks out on the incised stream loop and on the "island" it circumscribes. Beyond is the level surface of the Coconino Plateau, capped with resistant limestone of the Kaibab Formation, the same rock that caps the rims of Grand Canyon.

In the upper canyon walls, exposed as a series of ledges and slopes, the Kaibab Limestone is made up of thick, fairly massive layers of sandy or silty limestone and dolomite separated by thinner layers of limy, shaly siltstone and sandstone. (Dolomite differs from limestone in containing magnesium carbonate as well as calcium carbonate. When it weathers, it forms a rough surface prickly to the touch.) Many of the layers bear marine fossils: clams, snails, bryozoans, and brachiopods. The nature of the limestone and dolomite, their fossils, and the amounts of silt and sand within and between them, suggest a near-shore marine environment.

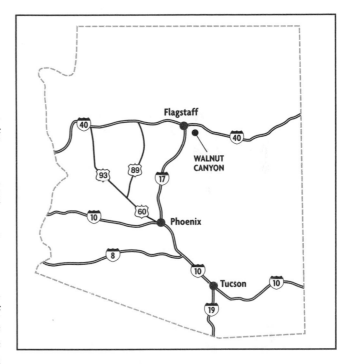

The lower canyon walls are formed of fine white cross-bedded sandstone, a sandy near-shore facies of the Toroweap Formation. This rock unit contains no fossils at Walnut Canyon, but displays patterns of cross-bedding like those of beach sandstone and coastal dunes. Like the Kaibab Formation of the upper canyon walls, the Toroweap Formation, mostly limestone in Grand Canyon, gradually becomes sandier toward the southeast as it approaches the Permian shoreline. Here it appears to be right along the strand line (see sidebar).

Many parallel vertical joints cut through the canyon walls, particularly in the cross-bedded Toroweap Formation. To some extent they control the drainage here, partly guiding, for example, Walnut Canyon's

RIGHT ABOUT FACIES

As we move from shore to sea, dune sand gives way to dry and then wet beach sand, and beach sand to sandy near-shore mud, which in turn yields, if we are good swimmers and capable divers, to fine calcium carbonate mud on the bottom of a warm, shallow sea, made largely of broken shells or, if we are in the tropics, fragmented reef corals.

Similarly, in sedimentary rocks, the features of a single formation may change from place to place where the formation is exposed. Recognizable in one location as cross-bedded dune or beach sandstone, elsewhere the same formation may be represented by limy mudstone, and still elsewhere it may become the pure limestone of shell beds and coral reefs.

Along with lithologic changes come biologic changes—changes in the types of animals and plants that inhabit different environments. And in geology we find that fossil assemblages change, too, as we move from shore to sea.

The sum of the lithologic characteristics (the nature of the rock) and the paleontologic characteristics (the nature of the fossils within it) are known as facies—and facies may change both horizontally, from place to place, as we have shown, or vertically, with time.

Exposures in Grand Canyon are ideal for studies of facies in Paleozoic sediments. For instance, exposures there show that the Kaibab and Toroweap Limestones each represent vertically a single transgression of the continent—each formation displaying a shallow-water facies overlain by a deeper-water facies and then a shallower-water facies again. Horizontally, too, the Kaibab and Toroweap Formations change facies from open-sea limestone in western Grand Canyon to a sandy, nearer-to-shore limestone facies at the east end of Grand Canyon.

In Walnut Canyon, well southeast of Grand Canyon, in both the Kaibab and Toroweap Formations, a near-shore facies predominates, with the Kaibab represented by very sandy limestone and the Toroweap showing the characteristic cross-bedding of beach sandstone. The varied and abundant assemblage of brachiopods, corals, bryozoans, and mollusks (clams, snails, and cephalopods) seen at Grand Canyon peters out as we approach Walnut Canyon, where corals are lacking and other fossils are small, mostly fragmented, and few and far between.

Still farther east the Toroweap becomes virtually indistinguishable from the windblown sands of the Coconino Sandstone, which clearly underlies it in Grand Canyon.

Small cliff dwellings peer from caves beneath the massive ledges of silty and sandy Kaibab Limestone. Limestone here was deposited nearer to shore than equivalent strata in Grand Canyon. Halka Chronic photo.

Cross-bedding in the Coconino Sandstone reflects its dune origin. Vertical joints, strongly developed here, to some extent guided development of Walnut Canyon. Halka Chronic photo.

sharp gooseneck bend. Steep ravines on the canyon walls are erosion-enlarged joints. Though most of the faults that cut across the canyon show only a few inches of displacement, one near the trail to the ruins has a displacement of about 10 feet (3 meters).

The trail drops down across ledges of Kaibab Formation limestone and dolomite—some of it sparkling with grains of fine quartz sand, some patched or completely coated with lichens. Fossil shellfish, particularly hollow molds left where actual shells were dissolved away by groundwater, can be seen in several of the ledges. Such fossils are studied by filling the little hollows with liquid latex, letting it dry, and then snapping the latex out of the hollows, giving perfect casts of the original shell. Salmon-colored geodes, dark brown limonite concretions, and nodules of hard white chert protrude from some limestone surfaces. The chert formed around sponges whose skeletons, made of tiny needles of silica, have left a fine tracery visible when the nodules are broken open.

Much of the trail around the "island" is just above the boundary between the Kaibab Formation and the Toroweap Formation. Unusually good exposures of the Coconino Formation, with its prominent cross-bedding, appear across the canyon.

GEOLOGIC HISTORY

Paleozoic Era. Most of the Paleozoic formations of Grand Canyon's walls underlie Walnut Canyon. But the smaller, shallower canyon cuts through only the uppermost layers, so we'll begin our story fairly late in

Permian time, when shallow seas twice crept in from the west. The two advances and retreats of the sea are well documented in Grand Canyon's upper walls, where the Toroweap and Kaibab Formations each involve a sequence of shoreline deposits of the advancing sea, marine limestones containing abundant fossils, and then shoreline deposits again. But at Walnut Canyon, near the farthest eastward extent of the Permian sea, the two formations each display evidence of a near-shore and beach environment, with sandstone and siltstone taking over from the limestone deposited in the more open sea north and west of here.

Mesozoic and Cenozoic Eras. At the end of Permian time the sea retreated and after a period of erosion, colorful Mesozoic strata—the Moenkopi, Chinle, and younger formations of the Painted Desert—were deposited on the eroded surface. Cenozoic sediments followed, including gravel, sand, and volcanic ash deposited in a large inland lake.

With uplift of the Colorado Plateau, erosion stripped away both Mesozoic and Cenozoic deposits near Walnut Canyon, though it left them in the Painted Desert area to the east. With the growth of the San Francisco Mountain volcano, streams that wound across the plateau surface were strengthened and able to entrench their former meandering courses, some of them perhaps following the routes of collapsed caverns in the limestone plateau.

Prehistoric people of the Sinagua culture moved into the canyon around 850 years ago, building one-story rooms with blocks of Kaibab Limestone mortared with clay. The move to Walnut Canyon seems to have been part of a population shift that followed the eruption of Sunset Crater, dated by tree-ring studies as having begun around A.D. 1100. Volcanic cinders from the eruption may have provided good mulch for Sinagua farms on the plateau surface, particularly around open "parks" and along shallow drainages, making them attractive to the agricultural people (see sidebar on page 143). Dwellings in Walnut Canyon were abandoned, though, before A.D. 1300, possibly because of a quarter century of drought between 1276 and 1299, a drought that also left its traces in tree rings in this area.

OTHER READING

Thybony, Scott,1996. *Walnut Canyon National Monument, On the Edge of the Past.* Western National Parks Association.

WUPATKI NATIONAL MONUMENT

ESTABLISHED: 1924
SIZE: 55 square miles (143 square kilometers)
ELEVATION: 4900 feet (1494 meters) at visitor center
ADDRESS: 6400 N. Hwy 89, Flagstaff, Arizona 86004

STAR FEATURES

- ✪ Prehistoric villages scattered along the edge of the Coconino Plateau, where Paleozoic sedimentary strata bend down eastward and then level out again beneath Mesozoic rocks of the Painted Desert.
- ✪ Views of the broad valley of the Little Colorado River, part of Arizona's Painted Desert, and of San Francisco Mountain, monarch of the San Francisco volcanic field.
- ✪ Evidence of volcanism and of the effect it may have had on prehistoric agricultural people.
- ✪ Unusual earthcracks that "breathe" in and out of underground crevices.
- ✪ Visitor center, museum, trails to the ruins (with guide leaflets), guided tours.

SETTING THE STAGE

West of the Little Colorado River and northeast of the San Francisco volcanic field, the Black Point Monocline defines the edge of the Coconino Plateau. Along

Moenkopi siltstone breaks along natural planes, providing good building stone. Halka Chronic photo.

it, Paleozoic sedimentary rocks bend eastward and plunge beneath Mesozoic rocks of the Painted Desert. Close to the monocline are ruins of several "apartment-house" villages dating back about 900 years, built by people of a predominantly Sinagua culture, possibly in response to a geologic event: the eruption of Sunset Crater.

Like other ruins in this area, Wupatki Pueblo was constructed with blocks of red siltstone from the Moenkopi Formation, broken along convenient natural joints. The pueblo rests on the brink of a large earthcrack, used for granaries and storage by the original inhabitants. Ruins of this village give an exciting glimpse into geology's influence on prehistoric people.

In the Wupatki area, the Kaibab Formation, easily recognized, hard, buff-colored Permian rock that surfaces much of the Coconino Plateau, dips fairly gently eastward. Here the formation shows near-shore traits: It is made up of silty and sandy limestone and limy siltstone and sandstone, rather than the purer limestone farther west (see sidebar on page 148). It contains pockets of marine fossils, mostly tiny clams and snails but also a few large, thick-shelled mollusks.

The Moenkopi Formation above it adds a deep red

Lomaki Pueblo rests on the brink of a large earthcrack. When the pueblo was occupied, dwellers built granaries and storage rooms within the earthcrack. But a small rock "dam" may have been constructed by Navajo to corral sheep. Halka Chronic photo.

color to the national monument area. This rock unit consists of red mudstone, siltstone, and sandstone, some thin-bedded and slabby, some thick-bedded and massive. Many layers are cross-bedded; others are marked with the tracery of mud cracks, ripple marks, and salt crystal impressions.

Several dark lava flows from the San Francisco volcanic field extend into and across the national monument, two of them reaching the Little Colorado River. A few purplish gray cinder cones are present in the monument as well, and the ground in many places is sprinkled with dark gray cinders from Sunset Crater (see Sunset Crater National Monument).

Stresses that developed as the Coconino Plateau rose and the Black Point Monocline took shape fractured both the Kaibab and the Moenkopi Formations. Joints are at almost right angles and mark the rock into rectangular slabs, a convenience for Sinagua stonemasons, who worked without metal tools. Most of the ruins here are constructed from bricklike slabs of Moenkopi siltstone and small boulders and cobbles of basalt.

Of special interest in this national monument are a number of large earthcracks in the Kaibab Limestone. Some of the cracks act like blowholes, with steady breezes blowing from them at times when air pressure outside is falling, and outside air blowing into them when air pressure is rising. Measurements of the amount of air "breathed" in and out show that these cracks communicate with extensive networks of underground cavities. Other cracks are partly filled with fine silt and clay, and often hold pools of rainwater. Some of the Sinagua villages were built near

Citadel's Sinagua pueblo builders blended irregular lumps of basalt with rectangular slabs of red sandstone. Halka Chronic photo.

such earthcracks, presumably because of the availability of water.

GEOLOGIC HISTORY

Paleozoic Era. Paleozoic history here resembles that of other Plateau areas: mostly marine sedimentation, with dunes forming in mid-Permian time. The only Paleozoic rock units appearing at the surface in the Wupatki area are the Kaibab Limestone and the Coconino Sandstone, both formed in Permian time in and near a shallow sea that came into this area from the west. Here the Kaibab Limestone is quite sandy, reflecting its nearness to the Permian shoreline. Its tiny fossil mollusk shells, or their impressions, may represent little creatures that clung to marine plants. Larger mollusks developed thick shells as protection against wave surge.

Mesozoic Era. Dark red siltstones and sandstones of the Moenkopi Formation are products of floodplains and coastal mudflats on a broad western continental slope. In several areas near the national monument, this formation contains fossils of large-bodied, flat-headed amphibians that swam or crawled about on Triassic floodplains and marshes. In places, tracks of a proto-dinosaur, *Chirotherium*, have been found, shaped like human handprints with the "thumb" on the outside instead of the inside, impressed into muddy surfaces and later covered with layers of sand. Where the mudstones have washed away, casts of the tracks appear on the undersides of overhanging sandstone ledges.

After the Moenkopi Formation was deposited, a thin but remarkably widespread layer of gravel, with rounded pebbles of hard Precambrian rock, sheeted across this area. Today the gravel makes up the Shinarump Conglomerate, a resistant unit that caps a number of mesas within the national monument. Younger Triassic rocks, some of them documenting volcanic explosions somewhere to windward, appear as the Chinle Formation in the Painted Desert to the east.

Cenozoic Era. Detailed studies of lava flows and cinder cones of the San Francisco volcanic field, and of terraces along the Little Colorado River, recount the Cenozoic history of this area.

After the establishment of the Colorado Plateau and its subdivisions, the Little Colorado River roughed out the wide valley we see here today. Volcanic eruptions west of this valley began about 6 million years ago as lava flows spread over the area. At times, earthquakes shook this region as faults displaced the plateau-capping basalts.

Periodically, lava flows from the volcanic field to the west intercepted the river, deflecting it from its course. Erosion by the Little Colorado River between 4 million and 2 million years ago, and eruption of the Black Point lava flow about 2.4 million years ago, created terraces 650 and 425 feet (200 and 130 meters) above the present river. Meanwhile, the great stratovolcano of San Francisco Mountain was slowly growing by additions of layer after layer of thick, silicic lava and volcanic ash. At its greatest size it may have been 16,000 feet high. It is now thought that a sideways explosion similar to that of the 1980 Mount St. Helens eruption left the horseshoe-shaped ring of peaks we

A sink near Citadel Ruin formed by solution along a fault in the Kaibab Limestone. Halka Chronic photo.

see today. During Pleistocene time, glaciers developed on the ring of peaks; a large one in its Inner Basin contributed outwash gravels that extend all the way to the Wupatki area.

Then, 150,000 to 80,000 years ago, renewed erosion reshaped the valley of the Little Colorado River and created terraces 80, 50, and 30 feet (25, 15, and 9 meters) above the present bed of the river. Since then, the Little Colorado River has cut down about 30 feet (9 meters) into the gravel deposits of the lowest terrace.

Agricultural Sinagua farmers came into this area about a thousand years ago. At first they built small pit houses near their farms. When Sunset Crater began to erupt around A.D. 1100 they probably fled, but they later returned, perhaps finding that the new layer of cinders scattered by the eruption served as a natural mulch, improving the productivity of their fields (see sidebar on page 143).

Pueblos were built at Wupatki, Wukoki, and other sites. Obtaining water from springs and seeps, earthcracks, and the Little Colorado, and catching and saving rainwater, the farming people remained here until about A.D. 1250. Then they abandoned their villages for reasons as yet uncertain—possibly because of depletion of soil and sources for firewood or because of prolonged drought. They may have merged with their pueblo-building relatives, the Hopis, who live northeast of here, or they may have migrated to the Verde Valley southwest of here, where the Verde River provided a more dependable water supply.

OTHER READING

Lamb, Susan, 1995. *Wupatki National Monument.* Southwest National Parks and Monuments Association.

Moore, R. B., and E. W. Wolfe, 1987. *Geologic Map of the East Part of the San Francisco Volcanic Field, North-Central Arizona.* USGS Miscellaneous Geologic Investigations Map MF-1960.

Thybony, Scott, 1987. *A Guide to Sunset Crater and Wupatki.* Southwest National Parks and Monuments Association.

ZION NATIONAL PARK

ESTABLISHED: 1909 as Mukuntuweap National Monument, 1919 as Zion National Park
SIZE: 229 square miles (593 square kilometers)
ELEVATION: 3666 to 8740 feet (1117 to 2664 meters)
ADDRESS: SR 9, Springdale, Utah 84767-1099

STAR FEATURES

- ✪ A dramatic canyon with towering walls of colorful Jurassic sandstone, relic of an ancient desert.
- ✪ The North Fork of the Virgin River, carver of the canyon—placid much of the year but often rising to a furious pitch during the summer rainy season.
- ✪ Smaller canyons, deep clefts between narrow rock ramparts, many controlled by parallel joints and faults.
- ✪ Stone arches, freestanding or still attached to the cliffs behind.
- ✪ Abundant evidence of rockfalls and slides, and a line of springs that contribute to them.
- ✪ Evidence of Quaternary volcanism.
- ✪ Visitor centers, museum, hiking trails (some with guide leaflets), guided walks, and evening programs. A geologic map of Zion is available at the visitor centers.

See frontispiece and color section for additional photographs.

SETTING THE STAGE

Deeply carved in the southern end of the Markagunt Plateau, Zion Canyon boasts some of the highest sheer cliffs in America, a few of them 2000 feet (600 meters) from top to base. The soaring monuments that confine the canyon, painted in delicate hues of pink and white, are shaped in a major Plateau scenery-maker, the Navajo Sandstone. The sweeping cross-bedding of this formation ornaments the canyon walls.

In Zion the Navajo Sandstone is 2000 feet (600 meters) thick; elsewhere it reaches 3000 feet (1000 meters). And it extends far beyond Zion, from central Wyoming (where it is called the Nugget Sandstone) to southern Nevada. Over all of this area it is remarkably similar in texture and composition, consisting of fine quartz sand with variable amounts of iron oxide, which give it its pink color, and calcite. Wherever its grains are cemented together with silica, the sandstone is

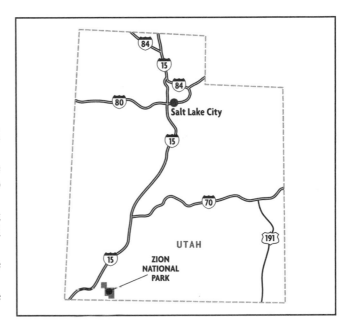

strong enough to stand in tall vertical cliffs.

What factors led to the building of this great body of sandstone? Its large-scale cross-bedding and uniformly fine, rounded, frosted sand grains show that it accumulated as dunes on a windswept desert. Its great aerial extent tells us that this desert once covered a region nearly as large as the Sahara. Horizontal siltstone layers between the slanting sets of cross-beds show us that silt and clay particles accumulated in flat interdune areas like those known from the Sahara and other modern deserts.

Gradual downcutting by the North Fork of the Virgin River shaped the Zion Canyon of today. Admittedly, the North Fork usually seems too clear, too placid to have accomplished all this erosion. But with summer thunderstorms or spring melting of snow on the Markagunt Plateau around its headwaters, the river picks up volume and speed, and pounds its bed with sand, pebbles, and even large boulders, continuing to cut down at an average rate of about 1.5 inches (3 centimeters) per century, roughly 1000 feet (305 meters) in the last million years. Fed by the more abundant snowfall and rainfall of Pleistocene time, downward cutting may have temporarily far exceeded this figure.

Canyon widening is the job of slides and rockfalls. The part of Zion Canyon that is accessible by road is

Sweeping cross-bedding in the Navajo Sandstone is a relic of a Jurassic desert as extensive as the modern Sahara. Tad Nichols photo.

edged with tumbled blocks and rockslide material. The Kayenta Formation, below the Navajo Sandstone, is responsible for most slides; its relatively soft siltstones and mudstones fairly invite erosion. As it wears away, the Navajo Sandstone is undermined, and despite its inherent strength eventually breaks away, sometimes along vertical joints, sometimes in large curving slabs.

Proof of the Kayenta Formation's role in canyon widening can be found in the Narrows, beyond the end of the paved trail. There, where the river has not yet cut down to the Kayenta Formation, Navajo Sandstone cliffs come right to the water's edge. But just below the Narrows, near the end of the road, the Kayenta Formation occurs low down on the canyon walls. A recent rockslide partly dams the stream, its giant boulders giving evidence that canyon widening is taking place.

Undermining and collapse are abetted by a line of springs along the contact between the Navajo Sandstone and the Kayenta Formation. Rainwater and snowmelt that filter down through the permeable sandstone eventually meet impermeable mudstones of the Kayenta Formation, and are turned aside along the contact between the two formations. Where the contact is exposed, the water emerges as a line of springs low down on the canyon walls.

The spring line at the base of the Navajo Sandstone, as well as smaller ones along interdune deposits, are also responsible for some of Zion's great arches. (Unlike those of Arches National Park, most of Zion's arches are not freestanding, but are large alcoves in the canyon walls.) Arches develop where there are relatively few vertical joints, where undermined Navajo Sandstone breaks away along curving fractures. Overhangs like that at Weeping Rock form the same way but on a smaller scale. Near springs, water in narrow cracks expands as it freezes on winter nights, helping to break the rock apart.

Where rock breaks along vertical joints, it also influences patterns of erosion. Some of the tributaries of the North Fork, as well as many tributaries of tributaries,

Contorted bedding such as that shown here originated in slumps or sand avalanches on ancient dunes. Tad Nichols photo.

follow the north-northwest trend of faults and joints that slash the strata here, giving an overall fabric to the land. The trend of these joints and faults parallels the trend of the Hurricane Fault, the major fault that separates the Colorado Plateau from the Basin and Range region to the west. Along the Hurricane Fault, offset is as great as 4000 feet (1200 meters). A further offset of up to 500 feet (150 meters) is provided by the East Cougar Mountain Fault, which cuts through eastern and northeastern parts of the park.

Details of weathering of the Navajo Sandstone are everywhere visible in the park. Wind etches the cross-bedding, desert varnish adds dark peacock tones to rock surfaces, black lichens stripe canyon walls, seeping water leaves white streaks of calcium carbonate, and growing plants pry rocks apart.

GEOLOGIC HISTORY

Mesozoic Era. Since Paleozoic rocks are exposed only in the northwest tip of the national park, where a tiny outcrop of Permian marine limestone comes to the surface along the Hurricane Fault, we'll begin our history with events of Mesozoic time.

As the Triassic Period began, this part of North America rose above sea level as a broad, westward-sloping coastal plain. Rivers and streams wound across the plain, eroding the top of the Permian limestone and depositing in this area about 1000 feet (300 meters) of sand and silt, mud and clay of the Moenkopi Formation. The dark red color of this formation comes from oxidized iron-bearing minerals such as biotite and hornblende, which in turn came from ancient Precambrian gneiss and schist eroded from the Ancestral Rocky Mountains in Colorado and New Mexico—rocks that can be seen today in Colorado National Monument and Black Canyon of the Gunnison National Park. The Zion area, close to the Mesozoic coastline, was repeatedly submerged in the sea, and as a result the Moenkopi Formation in this area contains two thick beds of limestone, complete with fossils of marine shellfish, as well as layers of salt and gypsum left in isolated lagoons where sea water evaporated. Lying close to the equator in Triassic time, the region was definitely tropical: warm and humid.

In mid-Triassic time, the nature and probably the source of sediments changed. A thin but widespread sheet of pebbly gravel was deposited by sheet floods and braided streams flowing north from highlands in central Arizona. The resulting Shinarump Conglomerate forms the distinctive lowest part of the Chinle Formation. The upper part of the Chinle is made up of fine muds mixed with volcanic ash from volcanoes to the west and southwest. The land remained low and marshy, with sluggish, meandering streams, shallow ponds and lakes, and forests of trees such as those we see in Florida's Everglades today. The purple, green, and gray mudstones of the Chinle Formation contain fossil amphibians and reptiles as well as petrified conifers, cycads, ferns, and horsetails.

After another cycle of erosion, many thin layers of sandstone, siltstone, and mudstone, with only a little volcanic ash, were superimposed, making up the Moenave and Kayenta Formations. Dinosaurs left their tracks in some of the Kayenta Formation mudstones.

As North America slowly drifted westward and northward, it entered subtropical latitudes, then as now the site of major deserts. On a wide, windswept Jurassic plain, with rising mountains to the west intercepting moisture from the ocean, sand dunes gradually piled up in a sand sea that stretched from Arizona to Wyoming. The dunes became the Navajo Sandstone, which is about 2000 feet (600 meters) thick in the Zion area—an astounding quantity of almost pure quartz sand.

An unconformity bevels the top of the Navajo Sandstone, and a few thin layers of red siltstone and another but much thinner eolian sandstone make up the Temple Cap Formation. Still higher lies the Carmel Formation, with conglomerate, sandstone, silty limestone, and gypsum testifying to encroachment by a narrow sea between here and new mountains in California—the Sierra Nevada, born as the continent collided with a subcontinent along its western coast.

Much later, after another very long period of erosion, a Cretaceous sea crept into this area from the east. Cretaceous sediments are rare in Zion National Park, appearing only on the lava-protected summit of Horse Ranch Mountain in the northern part of the park, and consist of pebbly conglomerate, sandstone, and mudstone derived from mountains to the west. We know from areas just east of Zion that Cretaceous strata there included a thick shale-sandstone sequence deposited in the Cretaceous inland sea—the last sea to invade the continent—and that these deposits were superceded by a blanket of Tertiary continental sediments washed westward off the Rocky Mountains. The Rockies, stretching in a great arc from Canada into New Mexico, developed during the Laramide Orogeny, in late Cretaceous and early Eocene time, about 72 million to 40 million years ago. The Colorado Plateau, with mountains developing to the east, west, north, and south, remained an island of relative serenity, slightly bowed and buckled but with its rock layers still essentially horizontal.

Cenozoic Era. As the Rockies rose, deep intermountain basins in Utah and Wyoming received rock debris washed from them. Lake and river sediments deposited in one of these basins now show up as the Claron Formation in Bryce Canyon National Park and Cedar Breaks National Monument. In the Zion area all Tertiary deposits have eroded away.

Later, extension, or pulling apart, of the Basin and Range region to the west resulted in a series of faults along the edge of the Colorado Plateau. The extension

Rocky, tree-covered talus slopes below Zion's towering walls bear testimony to many rockfalls. The West Rim Trail, at the bottom of the photograph, climbs the talus slope below the cliffs. Red Arch Mountain, site of a fairly recent rockfall, is at upper left. Ray Strauss photo.

THE WALLS OF JERICHO?

Geologically speaking, Zion Canyon is an active place. In addition to ongoing downcutting by the river, canyon widening continues, with rockfalls and landslides playing leading roles.

Between 3600 and 8000 years ago, one or more massive rockslides occurred at the lower end of Zion Canyon, 1 mile (1.6 kilometers) above the bridge and the junction of the two highway routes into Zion. The slides blocked the flow of the North Fork of the Virgin River, damming a long, narrow lake that lasted for thousands of years. Remains of the rocky barrier still jut into the canyon as an unsorted jumble of rock fragments, many of them immense. Some remaining lake sediments form the level area around Zion Lodge.

Red Arch, near Refrigerator Canyon, fell in historic time, obliterating a cornfield and a spring belonging to a local farmer. And in 1995 a landslide downstream from Zion Lodge once more blocked the Virgin River, which then burst its dam and washed out a section of the road, temporarily stranding visitors and employees at the lodge.

started in the south and moved north, extending parts of Nevada and adjacent areas hundreds of miles. Faults separating the Plateau from the Basin and Range region include the Hurricane and Sevier Faults, east and west of Zion. Both of these faults were active within the last 2 to 3 million years. Lava flows and cinder cones along the west edge of the park show that branches of the Hurricane Fault go deep enough to tap the basalt magma of the Earth's mantle. The resulting volcanic rocks have been dated between 100,000 and 1.4 million years old.

The primary geologic process of Quaternary time, however, was (and is) erosion. Here it was augmented by Basin and Range downfaulting, uplift of the Colorado Plateau, and rerouting of the Colorado River into the newly opened Gulf of California, which for a time came within 130 miles of Zion (see Bryce Canyon National Park sidebar). Erosion was also boosted by heavy Ice Age precipitation of Pleistocene time. Rockfall debris in benches above the lower part of Zion Canyon and Coalpit Wash formed in Pleistocene time. Zion also contains evidence of Quaternary lakes, their waters dammed by lava flows or by rockslides within the canyon.

BEHIND THE SCENES

Angels Landing. The trail to Angels Landing offers unusual views of both upper and lower Zion Canyon. The North Fork is deflected from a southward course into a tight loop around Angels Landing and the Organ. There is also a good view of the bold north face of the Great White Throne, capped (as are parts of Zion's East and West Rims) with thin layers of the Temple Cap Formation. Details of the Navajo Sandstone, especially its dune-style cross-bedding, are visible along the trail. Tributary streams such as that in Refrigerator Canyon follow the north-northwest trend of parallel joints, a feature easy to observe from the summit.

Checkerboard Mesa. Here, erosion by wind and rain has creased the surface of the Navajo Sandstone. Rivulets have incised small gullies along vertical joints, while wind and blowing sand have etched the

sweeping cross-bedding of this rock. Checkerboarding seems to be limited to north-facing slopes. Like mesas north of the highway, Checkerboard Mesa is capped with striped layers of the Temple Cap and Carmel Formations. Several arches in the Navajo Sandstone can be seen north of the highway.

Coalpit Wash. This small valley contains Pleistocene lake deposits—clay, silt, and limestone—and broken volcanic rock that form a bench north of the lava flow that dammed the valley about half a million years ago. The lava is associated with two cinder cones, together called Crater Hill. The lake deposits contain a fossil camel track, as well as fossil pollen that shows the climate to have been moister in Pleistocene time than it is at present.

Court of the Patriarchs. Standing at the west side of Zion Canyon, the Patriarchs display the large-

Cross-bedding combines with vertical fractures to give an unusual appearance to Checkerboard Mesa. Tad Nichols photo.

scale cross-bedding and pink and white colors of the Navajo Sandstone. Joints trending north-northeast divide the Patriarchs and control the erosion of nearby rock clefts.

The perfectly flat canyon floor below the Patriarchs is underlain by lake deposits formed when a rockslide dammed the Virgin River.

East Rim Trail and Observation Point (3.6 miles or 5.8 kilometers). Once the only route from Zion Canyon to communities farther east, the trail up Echo Canyon passes through a narrow slot formed by merging potholes in the stream. Such potholes take shape as boulders and pebbles—natural grindstones—are whirled in slight depressions in the streambed. As walls between successive potholes were ground away, the stream rushed through the deep, twisting chasm.

Above this narrow passage the canyon opens out, and the trail is surrounded with cliffs and steep slopes of Navajo Sandstone. Cross-bedding in the Navajo Sandstone is particularly prominent here, with sets of long, slanting cross-beds reflecting changes in wind direction. Desert varnish, lichens, red stain washing down from the Temple Cap Formation, and many other geologic features can be seen here.

The pink color of the lower part of the Navajo Sandstone is caused by tiny particles of hematite (iron oxide) within the formation. The color may mark a stable groundwater level long before Zion Canyon was carved. Elsewhere the Navajo Sandstone varies from white to yellow to entirely pink.

The trail eventually climbs into the Temple Cap and Carmel Formations above the Navajo Sandstone. Following the ledge at the top of the Navajo Sandstone, the trail reaches Observation Point, with a magnificent view of the Great White Throne, Angels Landing, and almost all of Zion Canyon.

Gateway to the Narrows. In upper parts of Zion, the canyon of the North Fork becomes so narrow that the river occupies its entire width. The Kayenta Formation is well below river level, and cliffs of Navajo Sandstone rise sheer and tall from the water's edge. Summer floods rushing through the Narrows fill the gorge from wall to wall.

At the Gateway to the Narrows, where the Kayenta-Navajo contact is just above river level, springs along the contact turn woodlands into swamps. Two recent rockslides here partly dammed the river, tumbling giant boulders into its course. Much of the rockslide is a mixture of sand and mud—broken-up rock—but plenty of larger rock fragments remain. The scar from which the slides came can be seen well up on the canyon wall.

Elsewhere the cliffs above the river are marked with streaks of black lichens, iridescent coats of desert varnish, and patches of white calcium carbonate leached from the rock itself. High-up red streaking is fine mud from the Temple Cap Formation, which over-

The view from the West Rim reveals the flat upper surface of the Markagunt Plateau, part of it forming the top of the Great White Throne (left half of photograph). Ray Strauss photo.

lies the Navajo Sandstone. Note the joint patterns in the Navajo Sandstone, and the tendency of this unit to break away in arches.

Boulders in the stream are representative of the country around this river's headwaters. Some are of dark gray basalt, the rock that caps most of the Markagunt Plateau. Across the river are several free-standing pinnacles of Navajo Sandstone, perched precariously on pedestals of Kayenta Formation siltstone that eventually will be worn away by the river.

Hanging Gardens. Along the line of the Navajo-Kayenta contact, and on a smaller scale along horizontal interdune deposits, spring water seeps from the Navajo Sandstone. The water's source is in rainfall and snowmelt on the surface of the Markagunt Plateau. Trickling downward through the porous sandstone, water is forced to flow sideways by relatively impervious mudstones of the Kayenta Formation or by similarly impervious interdune deposits. Wherever it then intercepts the canyon, it comes forth in springs that nourish hanging gardens of water-loving plants.

The steadily seeping water weakens surrounding rock by repeatedly freezing in winter and by dissolving the cement that holds sand and silt grains together. As a result, springs are commonly shadowed by overhanging ledges.

Hidden Canyon. This narrow ravine is one of the many deep, narrow, almost inaccessible canyons that cut the southern Markagunt Plateau, deep clefts eroded along joints trending north-northwest. Hidden Canyon

Zion's trails lead through narrow canyons and rugged uplands of bare rock, where the Navajo Sandstone is the major scenery-maker. Ray Strauss photo.

is about 1 mile (1.5 kilometers) long and at the most only 65 feet (20 meters) wide. Vertical or overhanging walls of Navajo Sandstone are marked with lichens and stained with desert varnish. Here and there are little grottos, the work of water and wind. A small natural bridge is present within the canyon as well.

Kolob Canyons. The unfrequented northwestern part of Zion National Park is no less attractive than Zion Canyon. The road follows Taylor Creek into this part of the park, traveling on Triassic rocks of the Moenkopi, Chinle, and Moenave Formations. Turning southeast across a landslide, it leaves Taylor Creek and crosses a thrust fault where Jurassic rocks have pushed westward over the Triassic units.

Except near the South Fork of Taylor Creek, the road then remains on the Moenave Formation. Views eastward show the long buttresses of the Kolob Plateau, walled with Navajo Sandstone and capped with

In the Kolob section of Zion National Park, trails wind into a wilderness of narrow canyons and steep coral-colored cliffs. Ray Strauss photo.

reddish ledges of the Temple Cap and Carmel Formations. The buttresses are separated by the Finger Canyons of the Kolob Plateau. Several trails wind into this wilderness, some leading to rock arches, among them Kolob Arch, its 310-foot (94-meter) span one of the longest known natural spans in the world. In dry weather the sandy floors of the Kolob canyons make enjoyable pathways, too.

Kolob Reservoir Road. Entering the national park on lava flows that mark its western margin, this road provides easy access to some of Zion's volcanic features. The magma of the lava flows and cinder cones rose along West and East Cougar Mountain Faults; their basalt composition tells us that these faults extend down through the Earth's crust to the mantle. The molten rock flowed down a former valley of North Creek and its tributary Grapevine Wash, displacing the streams. As surrounding rocks wore away, the resistant lava remained as a ridge.

Traveling north between the two Cougar Mountain Faults, where flat-lying rocks tilt eastward in response to the fault movement, the road passes rows of small domes, the larger Tabernacle Dome, and hoodoos and cliffs of white Navajo Sandstone.

Spendlove and Firepit Knolls are cinder cones, with craters at their summits. From the rim of Firepit Knoll—a short hike—vistas extend southward to the great cliffs of Zion Canyon, northward to the ragged edges of the Kolob Plateau, and eastward over more lava flows to parts of the Markagunt Plateau that surround the headwaters of the North Fork. To the west are the Basin and Range deserts of southwestern Utah. The East Cougar Mountain Fault runs between the Spendlove and Firepit cinder cones.

Petrified Forest. In the Chinle Formation in the southwest corner of the park, accessible by trail, Zion's Petrified Forest contains fossil logs similar to those in Arizona's Petrified Forest, which is in the same formation. Many of the logs occur in the Shinarump Conglomerate at the base of this formation. In places, cavities within the logs contain carnotite, a bright yellow uranium ore.

Weeping Rock. Rain and snowmelt on the Markagunt Plateau seep down through younger rock units and into the porous Navajo Sandstone. Prevented from further downward movement by fine-grained Kayenta Formation mudstone and siltstone, they eventually emerge as springs along the contact between the two formations. As elsewhere along this spring line, water-loving plants ornament the site. Because the water weakens the rock, making it more susceptible to erosion, some of the rock has fallen away here, leaving an arching overhang.

Zion–Mount Carmel Highway and Tunnel. Completed in 1930, the highway travels up Pine Creek Canyon and tunnels through the otherwise insurmountable cliffs of Navajo Sandstone that confine

The highway to Mount Carmel climbs a landslide before disappearing into a tunnel for the final ascent through the Navajo Sandstone. Ray Strauss photo.

Zion Canyon.

In this part of Zion the Moenave and Kayenta Formations form steep, ledgy slopes below the Navajo Sandstone cliffs. The highway zigzags up this slope and over a large but seemingly stable landslide (with views of the Great Arch on the opposite canyon wall), to the tunnel's western portal, right at the Kayenta-Navajo contact. A short trail near the east portal leads to a viewpoint and panoramic vistas of Pine Creek Canyon and lower Zion Canyon.

East of the tunnel, the highway winds through bare rock outcrops of Navajo Sandstone. Many side canyons parallel the north-northwest joint pattern prevalent in this area, well revealed on Checkerboard Mesa.

OTHER READING

Crawford, J. L., 1988. *Zion National Park: Towers of Stone.* Zion Natural History Association.

Eardly, A. J., and J. W. Schaak, 1991. *Zion: The Story Behind the Scenery.* KC Publications.

Geologic Map of Zion National Park. 1995. Zion Natural History Association.

Hamilton, Wayne, 1995. *The Sculpting of Zion.* Zion Natural History Association.

Warneke, Al, 1993. *An Introduction to the Geology of Zion National Park.* Zion Natural History Association.

GLOSSARY

aa lava—extremely rough, fragmented lava.

agate—a banded or otherwise decorative variety of quartz.

alluvial—deposited by rivers and streams.

amethyst—a lavender variety of quartz.

ammonite—an extinct, shelled relative of the modern squid, octopus, and chambered nautilus.

anticline—a fold in layered rocks that is convex upward.

arch—a stone arch formed by erosion of rock, not bridging a watercourse.

archaeology—the study of prehistoric people and their cultures.

arroyo—a gully or small canyon, dry most of the time.

ash—see **volcanic ash.**

badlands—rough, gullied topography in arid and semiarid regions eroded by infrequent but heavy rains.

basalt—dark gray to black volcanic rock poor in silica and rich in iron and magnesium minerals.

basin—a downwarped or downdropped area filled with sediment eroded from surrounding higher areas.

bedding—layering of sedimentary rocks.

bedrock—solid rock exposed at or near the surface.

bench—a relatively gentle slope partway up a steeper canyon wall.

bentonite—a soft, porous, light-colored rock formed by decomposition of volcanic ash.

biotite—black mica.

bomb, volcanic—a fragment of molten or semimolten rock thrown from a volcano.

boulder—a large, rounded rock fragment with a diameter greater than 10 inches (25 centimeters), usually transported by running water.

brachiopod—a marine shellfish with two bilaterally symmetrical shells, common as fossils in Paleozoic rocks.

breccia (rhymes with "betcha")—rock consisting of coarse, broken rock fragments imbedded in finer material such as volcanic ash.

bryozoan—a group of branching, coral-like invertebrates belonging to the phylum Bryozoa.

butte—a steep-walled hill capped with resistant rock.

calcite (calcium carbonate)—$CaCO_3$, the major mineral component of limestone and travertine,

also occurring as cementing material in sandstone and siltstone.

caliche (ca-LEE-chee)—whitish calcium carbonate–cemented gravel found on or near the surface in arid and semiarid climates, deposited as mineral-bearing water is drawn to the surface and evaporated.

caprock—a resistant rock forming the top of a butte, mesa, or plateau.

cavern—a large and usually complex cave.

cephalopod—a group of marine mollusks with tentacles, including octopuses, squid, and the chambered nautilus as well as extinct ammonites.

chert—a hard, dense form of silica that usually occurs as nodules in limestone.

cinder cone—a small, conical volcano built primarily of loose fragments of bubbly volcanic material thrown from a volcanic vent.

cinders—bubbly, popcornlike volcanic material.

clay—very fine rock material, with particles less than 0.00016 inch (4 microns) in diameter.

claystone—sedimentary rock formed of clay-sized particles.

cobble—a rounded rock fragment having a diameter of 2.5 to 10 inches (6.4 to 25 centimeters).

composite volcano—see **stratovolcano.**

conchoidal fracture—a smoothly curved fracture such as that occurring in broken glass.

concretion—a hard rounded mass of mineral matter accumulating in sedimentary rock.

conglomerate—rock composed of rounded, water-worn fragments of older rock.

contact—the boundary between two rock units.

continental—sedimentary rocks deposited on land or in lakes, by streams, wind, or ice.

continental crust—the part of the Earth's crust that forms the continents, including the continental shelves.

coral—a group of sea-dwelling animals that may deposit calcium carbonate in large reeflike masses.

core, Earth's—central part of the Earth, probably consisting of material high in iron and nickel.

crater—the funnel-shaped hollow at the summit of a volcano, including the vent from which lava and volcanic ash are ejected.

crinoid—a marine animal related to starfish, having a lilylike body on a long, segmented stem.

cross-bedding—slanting laminae within a sedimentary rock layer.

crust—the outermost shell of the Earth, above the mantle and core.

cuesta—a ridge with a long, gentle slope formed by a tilted, resistant rock layer, and a short, steep slope on the eroded edges of that and lower layers.

cyclic—occurring in cycles.

dendritic drainage—a treelike pattern of branching streams and rivulets.

desert pavement—a natural concentration of closely packed pebbles, the result of winnowing by wind action and rain.

desert varnish—a thin, glossy coating of dark brown or blue-black manganese and iron minerals on rocks in desert regions.

differential erosion—irregular erosion caused by differences in rock hardness or resistance.

dike—a sheetlike intrusion that cuts vertically or nearly vertically across other rocks. In igneous rocks, dikes are often called veins.

dip—the direction and degree of tilt of sedimentary layers, measured downward from horizontal. Also used as a verb.

displacement—offset along a fault or monocline.

dissected—cut into by stream erosion.

dolomite—a sedimentary rock consisting of calcium and magnesium carbonates.

dome—a more or less circular anticline in which rocks dip away in all directions. (See also **lava dome.**)

drag—bending of rock layers near faults, caused by friction along the fault surface.

earthcrack—an open crack caused by an earthquake.

entrenched—occupying a deep canyon cut by stream or river erosion.

eolian—caused by wind.

epidote—a yellowish or greenish mineral found in metamorphic rocks.

epoch—a unit of geologic time, subdividing a period.

era—the largest unit of geologic time

erosion—the process by which rock is loosened, dissolved, and worn away.

evaporite—minerals left behind by evaporation of sea or lake water; includes salt, gypsum, potash, and anhydrite.

exfoliation—a process in which concentric crusts of rock break away from a rock surface. Also called **spalling.**

extrusive igneous rock—rock formed of magma that flows out on the surface and solidifies there. Also called **volcanic rock.**

fault—a rock fracture along which displacement has occurred.

fault block—a segment of the Earth's crust limited on two or more sides by faults.

feldspar—a group of common, light-colored, rock-forming minerals containing aluminum oxides and silica.

floodplain—nearly horizontal land adjacent to a river channel, with sand and gravel layers deposited by the river during floods.

fold—a curve or bend in rock strata.

formation—a mappable unit of stratified rock.

fossil—remains or traces of a plant or animal preserved in rock; also, long-preserved inorganic structures such as fossil ripple marks.

frost heaving—lifting of rocks or soil by crystal expansion as water in them freezes.

frost wedging—prying apart of rock by crystal expansion as water freezes repeatedly in cracks and crevices.

fumarole—a vent through which volcanic gases or vapors are emitted.

geode—a small, hollow, somewhat spherical rock, often with crystals lining its inside wall.

glaciation—presence of or erosion by glaciers.

glacier—a large mass of ice driven by its own weight to move slowly downslope or outward from a center.

glauconite—a green mineral containing iron, thought to form in agitated sea water.

gneiss (pronounced "nice")—banded metamorphic rock formed from granite (which it commonly resembles), sandstone, and other continental rocks.

gooseneck—a curve of an entrenched stream or river course.

graben—a linear valley dropped down between two parallel or nearly parallel normal faults.

gradient—the angle of slope of a stream or river course.

granite—a granular intrusive rock composed primarily of crystals of quartz and feldspar, peppered with dark biotite or hornblende crystals; any light-colored, granular intrusive rock.

gravel—a mixture of pebbles, boulders, and sand not yet consolidated into rock.

groundwater—subsurface water, as distinct from rivers, streams, seas, and lakes.

group—a major rock unit consisting of two or more formations having certain characteristics in common.

gypsum—a common evaporite mineral, calcium sulfate; $CaSO_4 \cdot 2H_2O$.

hardpan—a relatively hard or impervious soil layer in which soil particles are cemented together by minerals added by soil water.

headward erosion—stream erosion in an upstream direction, particularly at the head of a gully or canyon.

hematite—a common dark red, iron oxide mineral; Fe_2O_3.

hogback—a steep, sharp-crested ridge with approximately equal slopes, one formed of a hard, steeply tilted caprock, the other of the eroded edges of the caprock and layers below.

honeycomb weathering—weathering of sandstone into small, deep pits, usually by wind.

hoodoo—a grotesque pinnacle or pillar of rock.

hornblende—a black or dark green mineral whose rodlike crystals are common in igneous rocks.

humus—decomposed plant material in soil.

Ice Ages—a period of glacial activity, most frequently referring to the Pleistocene Epoch. (See geologic calendar, p. 26.)

igneous rock—any rock formed from molten magma.

impermeable or **impervious**—not allowing penetration of water or other fluids.

interdune deposit—a silty or clayey deposit formed in a low, flat area between sand dunes.

intrusion—an igneous rock mass formed from molten magma that does not reach the surface.

intrusive igneous rock—rock created as molten magma intrudes preexisting rocks and cools without reaching the surface.

iridium—a chemical element rare in surface rocks, more common in meteorites and the Earth's mantle.

jasper—a dense, opaque, often colorful variety of chert, a variety of quartz.

joint—a rock fracture along which no significant movement has taken place.

karst—a distinctive type of rough, irregular landscape formed by solution of limestone.

laccolith—a dome formed by forceful injection of magma between sedimentary layers, doming upper layers.

lagoon—quiet near-shore water sheltered by a reef or offshore bar.

laminae—thin layers of rock visible on eroded rock surfaces.

landforms—recognizable features of the Earth's surface, such as hills, valleys, cliffs, mesas, and pinnacles.

Laramide Orogeny—mountain building of late Cretaceous and early Tertiary time, creating the Rocky Mountains.

lava—molten magma that has reached the Earth's surface, or the rock formed when such magma cools.

lava dome—a type of volcano characterized by magma that piles into a rounded dome above its conduit.

lava flow—an outpouring of molten lava or a rock mass formed from it.

lichen—a plant community consisting of a fungus and an alga, appearing as a flat, often circular crust.

liesegang rings—concentric rings or bands formed as minerals precipitate in water-saturated rock.

lime—a term commonly, though incorrectly, used for calcium carbonate.

limestone—a sedimentary rock consisting largely of calcium carbonate.

limonite—a yellow-brown iron oxide mineral. $2Fe_2O_3 \cdot 3H_2O$.

lithosphere—the outermost rocky layer of the Earth.

magma—molten rock. When extruded onto the Earth's surface, magma is usually called **lava.**

magma chamber—a reservoir of magma from which volcanic materials are derived, usually only a few kilometers below the surface.

magnetite—a black iron mineral having magnetic properties; $(Fe,Mg)Fe_2O_4$.

mantle—the thick, partly molten zone between the Earth's core and crust.

marine—formed in the sea.

meander—a loop on the sinuous course of a river.

mesa—a large flat-topped hill with a resistant caprock and steep slopes, larger than a butte but smaller than a plateau.

metabolic—having to do with life processes.

metamorphic rock—rock derived from preexisting rocks altered by heat, pressure, and other processes at depths well below the surface.

metamorphism—alteration of rock by heat, pressure, and chemical processes.

mica—a group of complex silicate minerals characterized by shiny, closely spaced, parallel layers that split apart easily.

mid-ocean ridge—a continuous, linear mountain range extending through the Earth's ocean basins, site of seafloor spreading.

mineral—a naturally occurring inorganic substance with a characteristic chemical composition, color, texture, and crystal form.

mollusk—an animal group including clams, snails, octopuses, squid, and several extinct forms.

monocline—a fold in otherwise horizontal sedimentary rocks that flexes in one direction only, with similar layers at different levels on either side.

mud cracks—shrinkage cracks in drying mud.

mudflow—a flowing mass of fine mud.

mudstone—rock formed from mud, with clay- and silt-sized particles.

natural bridge—a freestanding rock span created by a stream eroding and finally penetrating a rock spur.

normal fault—a fault in which the hanging (upper) wall moves downward relative to the footwall.

oceanic crust—part of the Earth's crust that occupies the seafloor and that originated along mid-ocean ridges.

outcrop—bedrock that appears at the surface.

overburden—rock material overlying a rock unit or mineral deposit.

pahoehoe lava—lava with a glassy, smooth, or ropy surface.

paleomagnetic dating—dating of rocks by comparison with known patterns of reversal of the Earth's magnetic field.

pebble—a rock fragment, commonly rounded, 0.2 to 2.5 inches (0.4 to 6.4 centimeters) in diameter.

pegmatite—exceptionally coarse-grained igneous rock found as dikes or veins in large igneous intrusions.

peneplain—a level or almost level erosion-caused surface of fairly wide extent.

period—a subdivision of geologic time shorter than an era, longer than an epoch.

permeable—allowing penetration by water or other fluids.

petrified—turned to stone.

petroglyph—rock carving.

phytosaur—an extinct swimming reptile similar to the crocodile.

pinnacle—a fingerlike tower of rock.

plate—a block of the Earth's crust, separated from other blocks by mid-ocean ridges, trenches, or collision zones.

plateau—a flat-topped tableland more extensive than a mesa.

Plate Tectonics, Theory of—a theory that explains the Earth's crust as divided into oceanic and continental plates, with new crust created at mid-oceanic ridges and old crust subducted and remelted along some continental margins.

pothole—a circular depression ground out by pebbles, cobbles, and sand swirled by running water, or formed by solution in small pools of rainwater.

pressure ridge—a transverse ridge in a lava flow, caused by pressure of continued flow.

pumice—light-colored, frothy, lightweight volcanic rock.

quartz—crystalline silica, SiO_2, a common rock-forming mineral.

quartzite—sandstone consisting chiefly of quartz grains welded firmly together with silica.

racetrack valley—a circular or oval valley around a mountain mass, shaped in the eroded edges of weak sedimentary rock layers that formerly arched over the mountain.

radiometric dating—dating of rocks by analysis of decay of radioactive minerals.

raindrop imprints—small, circular impressions formed by rain falling on soft sediment.

redbeds—predominantly reddish sedimentary rocks.

reverse fault—a fault in which the hanging (overhanging) wall moves upward relative to the footwall.

rift valley—a deeply faulted valley; a separation in the Earth's crust.

rimrock—resistant rock forming canyon or plateau rims.

ripple mark—a pattern of small ridges formed as water or wind transports and deposits sand or silt.

rockslide—a landslide involving a large proportion of rock.

salt anticline—an anticline pushed up by a rising mass of salt.

salt dome—a dome pushed up by a rising mass of salt.

sand—rock fragments 0.0025–0.016 inch (62–4000 microns) in diameter.

sandstone—sedimentary rock formed of sand-sized particles.

schist—crystalline metamorphic rock that tends to split along parallel planes.

scour marks—striations eroded in a rock face by streams or slides in an adjacent gully.

seafloor spreading—movement of oceanic crust away from mid-ocean ridges, and creation of new oceanic crust at the ridges.

sediment—fragmented rock, as well as shells and other animal and plant material, deposited by wind, water, or ice.

sedimentary rock—rock composed of particles of other rock transported and deposited by water, wind, or ice.

selenite—a clear variety of gypsum commonly found in veins.

shale—fine-grained mudstone or claystone that splits easily along bedding planes.

shield volcano—a low dome-shaped volcano formed from fluid lava such as basalt.

silica—a hard, resistant mineral—SiO_2—that in its crystal form is quartz. It also occurs as opal, chalcedony, siliceous sinter, chert, and flint.

silicic—having a high proportion of silica.

sill—a flat igneous intrusion that has pushed between layers of stratified rock.

silt—rock fragments 0.00016–0.0025 inch (4–62 microns) in diameter.

siltstone—rock made of silt-sized particles.

sink or sinkhole—a more or less circular depression created when part of the roof of a cavern collapses.

slump—a landslide in which rock and earth slide as a single mass along a curved slip surface.

solution cavern—a cavern formed by solution of limestone.

solution valley—a valley formed by collapse of a solution cavern or by joining of a row of sinks.

spalling—peeling away of thin surface layers of rock.

spring line—a rock contact along which groundwater comes to the surface.

squeeze-up—a jagged mass of lava pushed up through a crack in an overlying lava crust.

strata—layers of sedimentary (and sometimes volcanic) rocks.

stratified—layered.

stratigraphic—pertaining to the study of strata (and therefore of sedimentary rocks).

stratovolcano—a volcanic mountain built of layers of lava, breccia, and volcanic ash.

stromatolite—fossil mound–forming calcareous algae.

structure—pertaining to the form and arrangement of rocks, in particular to faults and folds.

subduction—the downward plunge of one tectonic plate below another.

syncline—a fold that is convex downward.

talus—a mass of large rock fragments lying at the base of a cliff or steep slope from which they have fallen.

terrace—a horizontal shelf along a river valley, formed when a river erodes a deeper valley within its former floodplain.

travertine—a hard, dense limestone deposited by calcium carbonate–laden water of streams, hot springs, and caves.

tree-ring dating—a method of dating by comparing growth rings of trees or other wood.

trench—a seafloor depression marking the zone where one tectonic plate dives beneath another.

trilobite—a crablike invertebrate that became extinct at the end of Paleozoic time.

unconformity—a substantial break or gap in the geologic record, caused by an interruption of deposition, by uplift and erosion, or by igneous activity.

uplift—a large, high area produced by folding or faulting, as an arch, dome, or fault block.

vein—a thin, sheetlike intrusion into a crevice, often with associated mineral deposits. Veins may also be composed of minerals deposited by groundwater.

vent—any opening through which volcanic material is ejected.

volcanic ash—fine particles of pulverized magma exploded or blown from a volcano.

volcanic neck—solidified lava that cooled within a volcanic conduit, generally more resistant than surrounding rock.

volcanic rock—rock formed from magma exploded from or flowing out on the Earth's surface.

wash—a Southwestern term for a normally dry streambed.

weathering—changes in rock due to exposure to the atmosphere.

window—a hole through the rock, usually smaller than an arch or bridge, and well above surrounding surfaces.

INDEX

ABOUT THE AUTHORS

Halka Chronic, a geologist living in Sedona, Arizona, began her career at the Museum of Northern Arizona in Flagstaff. She later taught at the University of Michigan summer field camp and at Haile Selassie University in Addis Ababa, Ethiopia. She was for a time an editor-writer for the National Center for Atmospheric Research, and has consulted with the U.S. National Park Service in identifying and describing geologic landmarks in the Southern Rocky Mountain region. She has also been a consultant for groundwater and petroleum research. Living in Colorado for more than 30 years, she raised four daughters, two of them now geologists and her coauthors.

Her academic degrees include a B.A. in zoology from the University of Arizona, an M.A. in biology from Stanford University, and a Ph.D. in geology from Columbia University. Among her published books are *Roadside Geology of Colorado* (now in its second edition), *Roadside Geology of Arizona, Roadside Geology of Utah, Roadside Geology of New Mexico*, and three other volumes of the Pages of Stone series.

Growing up the daughter of two Ph.D.s in geology, young Lucy Chronic was more likely to hear discussions of plate tectonics than the prospects of the local ball team's chance of winning the pennant. Throughout her life, she has pursued interests in natural history, biology, and archaeology, but the family pull toward geology eventually won

out with her B.A. in geology from Carleton College and an M.S. in paleontology from the University of Wyoming.

Her professional life has remained eclectic, with writing and scientific interpretation being the threads that tie her various pursuits together. She has worked as an archaeologist, educational writer, scientific writer, and interpreter in state and national parks.

She resides in the mountains of central Idaho with her husband and two daughters, enjoying the wonders of our natural world.

THE MOUNTAINEERS, founded in 1906, is a non-profit outdoor activity and conservation club, whose mission is "to explore, study, preserve, and enjoy the natural beauty of the outdoors. . . . " Based in Seattle, Washington, the club is now the third-largest such organization in the United States, with seven branches throughout Washington State.

The Mountaineers sponsors both classes and year-round outdoor activities in the Pacific Northwest, which include hiking, mountain climbing, ski-touring, snowshoeing, bicycling, camping, kayaking and canoeing, nature study, sailing, and adventure travel. The club's conservation division supports environmental causes through educational activities, sponsoring legislation, and presenting informational programs. All club activities are led by skilled, experienced volunteers, who are dedicated to promoting safe and responsible enjoyment and preservation of the outdoors.

If you would like to participate in these organized outdoor activities or the club's programs, consider a membership in The Mountaineers. For information and an application, write or call The Mountaineers, Club Headquarters, 300 Third Avenue West, Seattle, Washington 98119; 206-284-6310.

The Mountaineers Books, an active, nonprofit publishing program of the club, produces guidebooks, instructional texts, historical works, natural history guides, and works on environmental conservation. All books produced by The Mountaineers fulfill the club's mission.

Send or call for our catalog of more than 450 outdoor titles:

The Mountaineers Books
1001 SW Klickitat Way, Suite 201
Seattle, WA 98134
800-553-4453
mbooks@mountaineers.org
www.mountaineersbooks.org

The Mountaineers Books is proud to be a corporate sponsor of Leave No Trace, whose mission is to promote and inspire responsible outdoor recreation through education, research, and partnerships. The Leave No Trace program is focused specifically on human-powered (non-motorized) recreation.

Leave No Trace strives to educate visitors about the nature of their recreational impacts, as well as offer techniques to prevent and minimize such impacts. Leave No Trace is best understood as an educational and ethical program, not as a set of rules and regulations.

For more information, visit *www.lnt.org*, or call 800-332-4100.

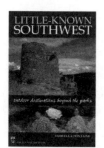